BIO-INORGANIC CHEMISTRY

ELLIS HORWOOD BOOKS IN INORGANIC CHEMISTRY

METAL IONS IN SOLUTION
J. BURGESS, Department of Chemistry, University of Leicester

BIO-INORGANIC CHEMISTRY
R. W. HAY, University of Stirling

METAL AND METALLOID AMIDES
M. F. LAPPERT, FRS, University of Sussex, A. R. SANGER, Alberta Research Council, R. C. SRIVASTAVA, University of Lucknow, India, and P. P. POWERS, Stanford University, California

SOLUTION EQUILIBRIA
F. R. HARTLEY, Royal Military College of Science, Shrivenham, C. BURGESS, Glaxo UK Limited, and R. M. ALCOCK, Severn-Trent Water Authority

INORGANIC SOLUTION CHEMISTRY
J. BURGESS, Department of Chemistry, University of Leicester (In preparation)

BIO-INORGANIC CHEMISTRY

R. W. HAY, B.Sc., Ph.D.
Reader in Chemistry
University of Stirling, Scotland

ELLIS HORWOOD LIMITED
Publishers · Chichester

Halsted Press: a division of
JOHN WILEY & SONS
New York · Chichester · Brisbane · Toronto

First published in 1984 by

ELLIS HORWOOD LIMITED
Market Cross House, Cooper Street, Chichester, West Sussex, PO19 1EB, England

The publisher's colophon is reproduced from James Gillison's drawing of the ancient Market Cross, Chichester.

Distributors:

Australia, New Zealand, South-east Asia:
Jacaranda-Wiley Ltd., Jacaranda Press,
JOHN WILEY & SONS INC.,
G.P.O. Box 859, Brisbane, Queensland 40001, Australia

Canada:
JOHN WILEY & SONS CANADA LIMITED
22 Worcester Road, Rexdale, Ontario, Canada.

Europe, Africa:
JOHN WILEY & SONS LIMITED
Baffins Lane, Chichester, West Sussex, England.

North and South America and the rest of the world:
Halsted Press: a division of
JOHN WILEY & SONS
605 Third Avenue, New York, N.Y. 10016, U.S.A.

British Library Cataloguing in Publication Data
Hay, R.W.
Bio-inorganic chemistry. –
(Ellis Horwood series in bio-inorganic chemistry)
1. Biological chemistry 2. Chemistry, Inorganic
I. Title
574.19'214 QH345

Library of Congress Card No. 84-4680

ISBN 0-85312-200-8 (Ellis Horwood Limited – Library Edn.)
ISBN 0-85312-766-2 (Ellis Horwood Limited – Student Edn.)
ISBN 0-470-20066-9 (Halsted Press)

Typeset in Press Roman by Ellis Horwood Limited.
Printed in Great Britain by The Camelot Press, Southampton.

Table of Contents

To
Alison
David, Caroline, Susan and Marion

Preface

Over the last decade or so there has been a growing awareness of the importance of a wide range of metallic and non-metallic elements in biological systems. Many journals covering inorganic chemistry now devote sections to bio-inorganic chemistry and it is likely that interest in this new area of inorganic chemistry will increase in coming years. A number of courses in inorganic chemistry now devote a series of lectures to the field of bio-inorganic chemistry. The present book should be of interest to undergraduates and post-graduate students taking courses in this field. The book should also be of value to research workers who would like an introduction to this new area of inorganic chemistry. A number of books and review articles are available on the more advanced aspects of the subject, and it is hoped that the present book will provide the basis for more advanced study in this field.

Bio-inorganic chemistry by its very nature is an interdisciplinary area, and as a result, readers may be well aquainted with the material in some sections of the book but less familiar with other sections. In order to provide material which will be of interest to inorganic chemists, biochemists and biologists, the coverage has been kept as broad as possible. References are not included in the text, but leading articles and reviews are summarized in the bibliography at the end of each chapter so that the interested reader can readily consult the original literature.

Finally, that this book ever reached completion is due to a number of people, to all of whom I owe my sincere thanks: Dr Mahesh Pujari who deciphered my handwiriting with remarkable patiencc, speed and accuracy in producing the complete typed manuscript; Ellis Horwood my publisher for his gentle cajoling as deadlines came and went, and finally my wife and family who tolerated my absences during the labours of authorship.

R. W. Hay
Chemistry Department,
University of Stirling, 1983

CHAPTER 1

General Background

GENERAL SURVEY

Bio-inorganic chemistry or inorganic biochemistry is concerned with the function of all metallic and most non-metallic elements in biology. It can be defined as the biochemistry of those elements whose chemistry normally constitutes the province of inorganic chemists. The scope of inorganic biochemistry is broad, stretching from chemical physics to clinical medicine. The subject has developed dramatically during this decade. The general upsurge of interest is connected with (1) improved analytical methods, (2) less time-consuming preparative techniques, (3) the successful application of spectroscopy and diffraction techniques, (4) the improved synthesis of simple inorganic complexes used to model or mimic various aspects of biological molecules, (5) the increased concern about the environmental hazards caused by some metal ions, (6) the use of metal ions or complexes as therapeutic agents, and (7) the recognition of the importance of an increasing number of trace elements in plant, animal and human nutrition.

METALS AND NON-METALS

Some 25 elements are currently thought to be essential to warm blooded animals. Ten can be classified as trace metal ions: Fe, Cu, Mn, Zn, Co, Mo, Cr, Sn, V and Ni, and four as bulk metal ions: Na, K, Mg and Ca. In addition there is some tentative evidence that Cd and Pb may be required at very low levels. The non-metallic elements are H, B, C, N, O, F, Si, P, S, Cl, Se and I. There is also evidence that Sn, As and Br may possibly be essential trace elements. Fig. 1.1 shows the periodic distribution of the elements essential to life. A noteworthy feature is that of the transition elements, only Mo is of importance in the second row.

PRIMORDIAL DEVELOPMENT

The earth is 4700 million years old and life is known to have existed for at least 3500 million years, most probably preceded by a period of chemical evolution

IA	IIA	IIIA	IVA	VA	VIA	VIAA	VIII	VIII	VIII	IB	IIB	IIIB	IVB	VB	VIB	VIIB	0
H																	He
Li	Be											B	C	N	O	F	Ne
No	Mg											A1	Si	P	S	Cl	Ar
K	Ca	Sc	Ti	V	Cr	Mn	Fe	Co	Ni	Cu	Zn	Ga	Ge	As	Se	Br	Kr
Rb	Sr	Y	Zr	Nb	Mo	Tc	Ru	Rd	Pd	Ag	Cd	In	Sr	Sb	Te	I	Xe
Cs	Ba	Ln	Hf	Ta	W	Re	Os	Ir	Pt	Au	Hg	Tl	Pb	Bi	Pc	At	Rn
Fr	Ra	Ac	Th	Pa	U												

Bulk biological elements. Trace elements believed to be essential for plants or animals.

Possibly essential trace elements.

Fig. 1.1 – The distribution of the elements essential for life.

lasting about 500 million years, (man has existed for 5-10 million years). During this period of chemical evolution it is generally believed that all of the molecules essential for life were successfully synthesized. Beginning with pure inorganic compounds, a set of inorganic-organic reactions took place leading to the formation of the amino acids and nucleotides, the precursors of proteins and nucleic acids. Similarly during this period of chemical evolution a mixture of inorganic and organic polyphosphates was synthesized. These compounds were required to provide the primary chemical reserves of free energy for further evolution. As a result, polymerization and organization occurred, and after a further period of development, the first cell evolved.

If this initial cell functioned in a manner similar to our present cells, a minimum of about 100 different protein molecules would be required to maintain protein synthesis and anaerobic energy production. Presumably incorporated into this cell were a series of metal ions. Some ions assumed functions required for the integrity of the cell, such as osmotic pressure, others (but initially very few) assumed functions of catalytic interest. The main metal ion catalyst (perhaps the only one), must have been the magnesium ion, as all the reactions involved in fundamental cell reactions, such as protein biosynthesis and anaerobic energy production, require Mg(II) ions. Magnesium ions are also known to catalyse several prebiotic condensation reactions.

Further support for the assumption that Mg(II) ions were important early in bio-evolution arises from its presence in seawater and sea sediment, an environment similar, it is believed, to that from which our cell system once evolved (Table 1.1). Another important biological process which requires magnesium is

Table 1.1 – Ion concentrations in sea water and extracellular blood plasma.

Ion	Sea water (mM)	Blood plasma (mM)
Na^+	470	138
Mg^{2+}	50	1
Ca^{2+}	10	3
K^+	10	4
Cl^-	55	100
HPO_4^{2-}	0.001	1.
SO_4^{2-}	28	1
Fe	0.0001	0.02
Zn^{2+}	0.0001	0.02
Cu^{2+}	0.001	0.015
Co^{2+}	$10^{-5.5}$	0.002
Ni^{2+}	10^{-6}	0

photosynthesis. In photosynthesis chlorophyll, a magnesium porphyrin derivative, captures light quanta and then utilizes this energy to fix carbon dioxide and to evolve oxygen. Once the porphyrins had been synthesized the abundance of Mg(II) ions would readily lead to the formation of chlorophyll.

AVAILABILITY OF Fe(II) AND Mn(II) UNDER REDUCING CONDITIONS

The atmosphere during the time of chemical evolution and of early life is generally supposed to have been a reducing one, characterized by a moderate partial pressure of methane (0.01 to 0.001 atm) and a very low oxygen pressure. Thermodynamic data for the system air-sea-sediment indicate that at equilibrium, the redox potentials in the seas were as low as −325 mV (pH 8.1), that is close to the hydrogen electrode (−480 mV at pH 8.1). Inorganic sulphur would be in the form FeS_2(s), the major excess of iron would be in the form Fe_3O_4(s), and there would be a certain amount in solution as Fe(II) ($\leqslant$ 0.2 mM Fe(II)). It is believed that the concentration of Mn(II) was much higher in primordial seas, perhaps as high as 50 mM. Thus these two metals would become of importance in early redox processes.

Some of the first metalloproteins appear to have been the iron-sulphur proteins. It has been suggested that inorganic electron carriers FeS_2(s) and $Fe_{0.86}S$(s) (pyrrhotite) were present in the area where the first proteins evolved. When polypeptides containing cysteine residues formed they reacted with the iron sulphides to give primitive iron-sulphur proteins. The relatively high concentrations of Fe(II) and Mn(II) in the oceans would lead to their incorporation into other ligands such as the porphyrins. This incorporation would lead to the haemoproteins, and in the case of manganese to components required for photosynthesis in the primitive blue-green algae. Some blue-green algae contain up to 12 atoms of Mn per reaction centre. These various components, along with chlorophyll, were necessary for photosynthesis which led to the release of oxygen into the atmosphere.

TRACE METALS, A SURVEY

In many cases the exact role of the trace elements is still not completely known. However, the great majority of these function as key components of essential enzyme systems or of other proteins (for example the haemoprotein hacmoglobin) which perform vital biochemical functions. The general roles of metal ions in biological processes are summarized in Table 1.2.

Zinc

Zinc was shown in 1934 to be essential for the normal growth and development of mammals, and is present to the extent of 1.4 to 2.3 g in the human body. Although zinc(II) is a d^{10} ion it has many similarities to the transition metal ions. In aqueous solution, oxidation states other than Zn(II) do not occur. Zn(II) acts predominantly as a Lewis acid, and is found in many metalloenzymes such as carboxypeptidase and carbonic anhydrase. Its role is considered in detail in Chapter 4. Zinc complexes are generally good buffers, and are used in pH control *in vivo*. In the body, there are about 18 zinc metalloenzymes and about 14 zinc-ion activated enzymes. Zinc(II) being a symmetrical d^{10} ion is fairly hard, and interacts strongly with oxygen and nitrogen donors.

Table 1.2 – General roles of metal ions in biological processes.

	Na, K	Mg, Ca (Mn)	Zn, Cd(Co)	Cu, Fe, Mo (Mn)
Type of complex	weak	moderate	strong	strong
Biological functions	charge transfer, nerves	trigger reactions, hydrolysis, phosphate transfer	hydrolysis, pH control	oxidation and reduction reactions
Ligand atom preferred	O	O	N and S	N and S

Iron

Iron is the most abundant transition metal in the human body, 4.2 to 6.1 g in the average man. In aqueous solution, Fe(II) and Fe(III) are the normal oxidation states, with Fe(II) being readily oxidized by air to iron(III) except in acidic solution. Both oxidation states are strong Lewis acids as evidenced by the pK_1 of iron(III) which is 2.2 ($pK_2 = 3.3$). The pK_a of $[Fe(H_2O)_6]^{2+}$ is about 9.5. Iron(III)

$$[Fe(H_2O)_6]^{3+} \overset{K_1}{\rightleftharpoons} [Fe(H_2O)_5OH]^{2+} + H^+$$

has an ionic radius of 0.67 Å and is a hard acid, while Fe(II) with a radius of 0.83 Å is borderline between the hard and soft acids. Dimeric iron(III) aquo-complexes are readily formed

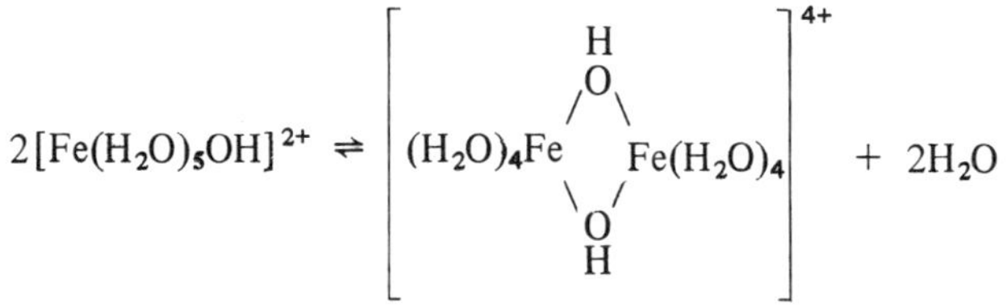

The addition of base causes further release of protons with linkage of cations and polymer formation. These polymeric iron complexes are of high molecular weight (*ca* 150, 000) and are roughly 70 Å in diameter. The iron storage protein ferritin consists of a protein shell surrounding a microcrystalline iron(III)-hydroxide phosphate sphere which bears a striking physical and chemical resemblance to the synthetic iron(III) oxo polymers.

As hydrolysis of iron(III) leads to polymer formation and eventual precipitation, iron(III) is essentially insoluble at neutral pH in distilled water (the solubility is calculated to be 10^{-18} M at pH 7). However, Fe(II) is soluble to the extent of 0.1 M at pH 7. Only by appropriate ligands is it possible to maintain soluble iron(III) in neutral and basic solutions. Appropriate ligands are EDTA, NTA and citrate. Iron may be present *in vivo* as Fe(II) or Fe(III) or may redox between the two (see Chapter 6), depending upon the ligand to which it is attached.

In haemoglobin and myoglobin (the oxygen binding proteins of blood and muscle) iron is present as Fe(II) in protoporphyrin IX (**1.1**) which is incorporated in a protein molecule (M = 17,000 in myoglobin).

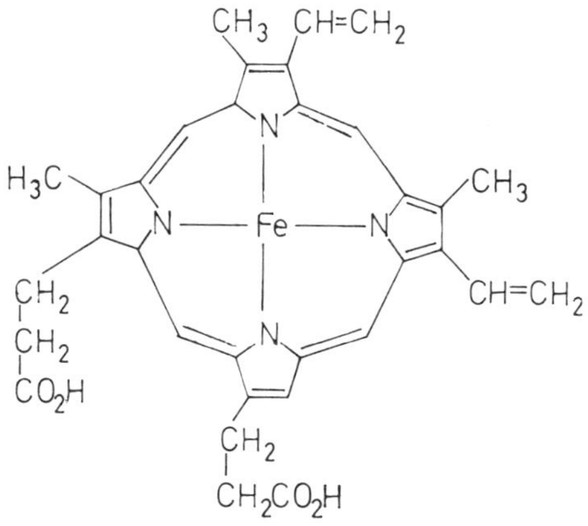

(1.1)

protoporphyrin IX

In the catalases, which are very large molecules (M = 247,500), and oxidases, iron is present as Fe(III). Catalase catalyses the decomposition of hydrogen peroxide:

$$2\,H_2O_2 \longrightarrow 2\,H_2O + O_2$$

For redox systems such as the cytochromes or ferredoxins the redox potential is tuned by the ligands present (Table 1.3).

Table 1.3 – Standard electrode potentials (E^0/V) for the Fe(III)/Fe(II) redox couple.

Ligand	E^0/V	Ligand	E^0/V
1,10-phen	+1.20	Cytochrome *f*	+0.4
2,2′bipy	+1.096	Cytochrome *c*	+0.25
Water	+0.77	Iron protoporphyrin	−0.12
Oxalate	−0.01	Horseradish peroxidase	−0.3
8-Hydroxyquinoline	−0.15		

Complexation of a transition metal ion results in significant changes in the electrode potential between the two oxidation states of the element:

$$\underset{\text{(Ox)}}{\text{Oxidized complex}} + Z\,\text{e} \rightleftharpoons \underset{\text{(Red)}}{\text{Reduced complex}}$$

The electrode potential is given by

$$E = E^0 + \frac{RT}{zF}\ln\frac{a\,(\text{Ox})}{a\,(\text{Red})}$$

where a (Ox) and a (Red) are the relative activities of the two forms and E^0 is the standard electrode potential, relating to unit relative activity, of each form at 298 K. The symbol z is the number of electrons transferred in the redox reaction. The standard free energy change ΔG^0 for the ion half reaction is related to E^0 by equation (1.1)

$$\Delta G^0 = -\,z\,FE^0 \tag{1.1}$$

The different magnitudes of the E^0 values in Table 1.3 reflect the different stabilities of the (II) and (III) oxidation states of iron in their various complexes.

The relationship between electrode potential and formation constant can be clarified by an example:

$$[Fe(H_2O)_6]^{3+} + e \longrightarrow [Fe(H_2O)_6]^{2+};\ \Delta G^0 = -\,nFE^0$$

$$[Fe(H_2O)_6]^{2+} + 3\ \text{phen} \longrightarrow [Fe(phen)_3]^{2+};\ \Delta G^0 = -\,2.303\,RT\,\lg\beta_3''$$

$$[Fe(phen)_3]^{3+} \longrightarrow [Fe(H_2O)_6]^{3+} + 3\ \text{phen};\quad \Delta G^0 = +\,2.303\,RT\,\lg\beta_3'''$$

where

$$\beta_3'' = \frac{\{Fe(phen)_3^{2+}\}}{\{Fe(H_2O)_6^{2+}\}\,\{phen\}^3} = 10^{21.3}$$

$$\beta_3''' = \frac{\{Fe\ (phen)_3^{3+}\}}{\{Fe(H_2O)_6^{3+}\}\,\{phen\}^3} = 10^{41.1}$$

and brackets indicate activities. By addition

$$[Fe(phen)_3]^{3+} + e \rightleftharpoons [Fe(phen)_3]^{2+}$$

and for this equilibrium

$$\begin{aligned} \Delta G^0 &= -FE_2^0 \\ &= -FE_1^0 + 2.303\,RT\,(\lg \beta_3''' - \lg \beta_3'') \end{aligned}$$

so that

$$E_2^0 = E_1^0 + \frac{2.303\,RT}{F}(\lg \beta_3'' - \lg \beta_3''')$$

As $E_1^0 = +\,0.77$ V

$$\begin{aligned} E_2^0 &= +\,0.77 + \frac{2.303 \times 8.314 \times 298(21.3 - 14.1)}{9.65 \times 10^4}\ \text{V} \\ &= 1.20\ \text{V} \end{aligned}$$

As with the bipyridyl ligand, the more positive value of E^0 (compared with the aquo system) indicates that the complexation by phen results in the greater stability of the lower oxidation state.

With oxalate and 8-hydroxyquinoline, coordination involves a negatively charged ligand with a positively charged cation. For such reactions, charge neutralization occurs in the complex,

$$Fe^{3+}\,(aq) + 3\ \text{oxime}^{-1} \rightleftharpoons Fe(\text{oxime})_3 + x\,H_2O$$

$$(\text{oxime}^{-1} = \text{anion of 8-hydroxyquinoline})$$

so that a good deal of solvating solvent is released (ΔS large and positive), leading to a large negative value of ΔG and to enhanced stability of the Fe(III) complex.

Copper

Copper is known in two oxidation states in solution Cu(I) and Cu(II). However, some Cu(II) complexes of tripeptides can be oxidized by air to Cu(III) species, and so Cu(III) may be of biological importance. The relative gain in crystal field stabilization energy for the change from d^9 Cu(II) to d^8 Cu(III) is an important factor in the overall thermodynamic stability of the Cu(III)-peptide complexes. The structure of copper(II) tetraglycine which contains three deprotonated amide groups is shown in (**1.2**). The Cu(III)/Cu(II) couple of this

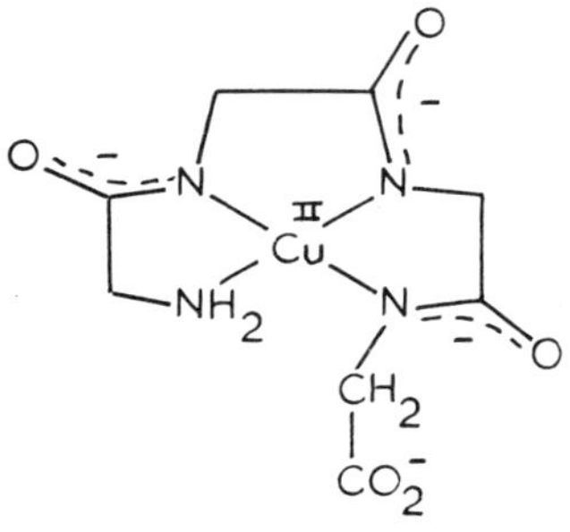

(1.2)

complex has a very low potential (0.63 V) so that O_2 oxidation to Cu(III) is thermodynamically possible. It would be expected that copper's main role *in vivo* would be in redox reactions and this is found to be the case. Copper is present in about 12 enzymes, whose functions range from the utilization of iron to the pigmentation of the skin. The redox potentials

$$Cu^{+}\,(aq) + e \longrightarrow Cu^{0};\quad E^{0} = 0.52\ \mathrm{V}$$

$$Cu^{2+}\,(aq) + e \longrightarrow Cu^{+}\,(aq);\quad E^{0} = 0.153\ \mathrm{V}$$

show that Cu(I) in aqueous solution disproportionates to Cu(II) and Cu(O)

$$2\,Cu^{+}\,(aq) \rightleftharpoons Cu^{0}(s) + Cu^{2+}\,(aq);\quad E^{0} = 0.37\ \mathrm{V} \qquad (1.2)$$

and $$K = \frac{[Cu^{2+}]\,[Cu]}{[Cu^{+}]^{2}} \approx 10^{6}$$

The instability of copper(I) in aqueous solution is partly due to the comparatively strong solvation of copper(II) in water. The thermodynamic stability of Cu(I) in solution can be considerably increased if solvents other than water, which do not enhance the solvation of copper(II) are considered. This effect is illustrated in Table 1.4. The equilibrium constant for eq. (1.2) decreases in

Table 1.4 – Electrode potentials (E^0/V) for the Cu(II)/Cu(I) couple.

Ligand	E^0/V	Ligand	E^0/V
laccase	+0.415	$(pyridine)_2$	+0.27
$(imidazole)_2$	+0.345	$(glycine)_2$	−0.16
blue proteins	+0.4	$(en)_2$	−0.38
cerruloplasmin	+0.39	$(alanine)_2$	−0.13
water	+0.16		

the order water > methanol > ethanol. In fact, Cu(I) does not disproportionate an acetonitrile, and copper(II) acts as a strong oxidizing agent in this medium.

The relative thermodynamic stabilities of Cu(I) and Cu(II) in aqueous solution is strongly dependent on the nature of ligands, Table 1.4, as we have previously seen with Fe(II) and Fe(III). Copper(I) is much softer than copper(II) therefore sulphur donors are bound more strongly by copper(I) than by copper(II), as are all unsaturated ligands such as *o*-phenanthroline and 2,2′-bipyridyl. The replacement of one then two water molecules by two nitrogen donors of imidazole (**1.3**) results in a more positive potential favouring Cu(I). The aliphatic α-amino acids and ethylenediamine lead to negative potentials and stabilize the harder Cu(II) (as $\Delta G^0 = -nFE$, the negative potentials give a positive free energy change for the reduction, which is therefore not favoured).

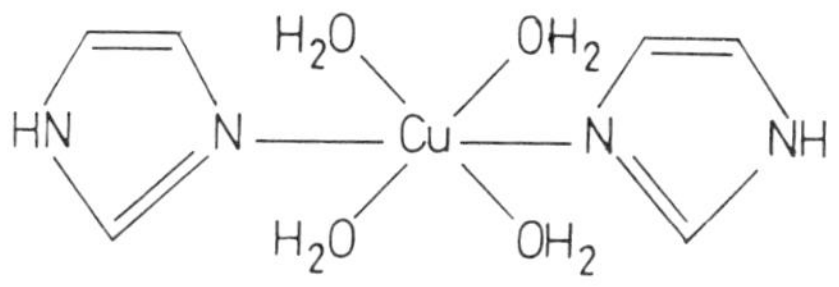

(**1.3**)

Copper(I) and copper(II) have very different stereochemistries. Copper(I) is often linear two coordinate, but a tetrahedral four-coordinate stereochemistry is common. In contrast, copper(II) with a d^9 configuration adopts a tetragonally distorted octahedral stereochemistry (as a result of the Jahn–Teller Effect).

One important principle which effects electron transfer reactions is the Franck–Condon principle which states that there must be no movement of nuclei during the period of the electronic transition. As a result, the geometry of the two species after electron transfer must be identical to that existing before the transfer occurred. Williams has suggested that the active metal site of a metalloenzyme is in a geometry approaching that of the transition state of the appropriate reaction and as such is uniquely fitted for catalytic action. Williams has used the term 'entatic state' to describe this situation [17].

The copper enzymes particularly involved in electron transfer reactions are the blue proteins (azurins) where the copper is in the copper(II) state (Table 1.5). One of the most unusual features of the blue proteins is the intensity of the blue colour. The absorption coefficients on a protein basis, for the band in the 600 nm region vary from 3500 $dm^3\ mol^{-1}\ cm^{-1}$ in azurin to 11,300 $dm^3\ mol^{-1}\ cm^{-1}$ for ceruloplasmin. Ceruloplasmin contains two 'blue' copper(II) ions per mole, giving an absorption coefficient of 5600 per atom blue copper(II).

Table 1.5 – Properties of some copper oxidases

Enzyme	Source	MW	Reaction catalyzed
Tyrosinase	Mushroom	120,000 4 Cu	Oxidation of monohydric phenols (e.g. cresol) to the *o*-dihydric compound and oxidation of the *o*-dihydric phenols (catechol) to the *o*-quinone
Laccase	Latex of the lac tree	120,000 4 Cu	*p*-Diphenols + $O_2 \rightarrow$ *p*-Quinones + H_2O
Ascorbic acid oxidase	Squash Cucumber	146,000 6 Cu	L-Ascorbate + $O_2 \rightarrow$ dehydroascorbate + H_2O
Ceruloplasmin	Human plasma	151,000 8 Cu	Oxidises *p*-phenylene-diamine, ascorbic acid and some *o*- and *p*-dihydroxy phenols
Pseudomones blue proteins (Azurins)	*Pseudomonas aeruginese*	16,400 1 Cu	Respiratory chain protein
Diamine oxidase	Pea seedlings	96,000 1 Cu 1 pyridoxal phosphate	$H_2N(CH_2)_nNH_2 + H_2O + O_2$ $\rightarrow NH_2(CH_2)_{n-1}CHO + NH_3$ $+ H_2O$

Human serum albumin has one specific copper(II) binding site. Copper bound to albumin is considered to be the transport form of copper(II) in blood. The copper binding site has been shown to involve the α-amino nitrogen of the NH_2-terminal aspartic acid residue, two intervening peptide nitrogens and the imidazole nitrogen of the histidine residue in position 3. The bonding site can be modelled using the tripeptide glycylglycyl-L-histidine which gives the complex **(1.4)** with copper(II). The dissociation constant of the Cu(II)-peptide complex is 1.2×10^{-16} compared with 6.6×10^{-17} for Cu(II)-albumin. The tripeptide glyglyhisNH_2 has been used for the treatment of Wilson's disease as it removes deposited copper in the brain and kidneys.

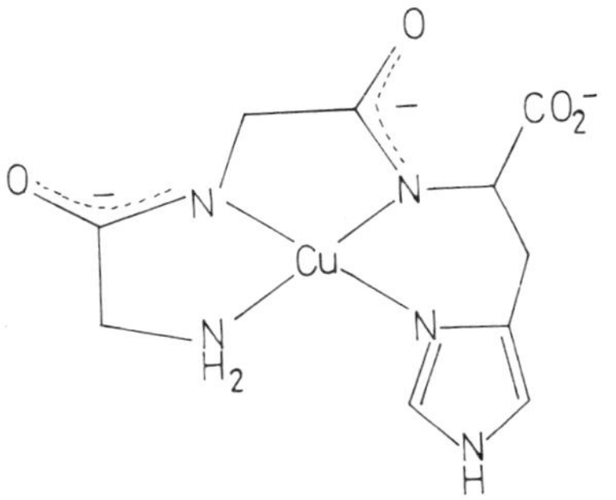

(1.4)

Mixed ligand complexes are used to transport copper(II) to serum albumin,

Cu^{2+} + amino acid anion ⇌ Cu^{2+}-amino acid complex

Cu^{2+}-amino acid complex + albumin ⇌ albumin-Cu^{2+}-amino acid complex

albumin-Cu^{2+}-amino acid complex ⇌ albumin–Cu^{2+} + amino acid anion

Molybdenum

Molybdenum is the only element of the second and third transition series known to be essential to life. Molybdenum is probably absorbed into living systems as the molybdate anion, MoO_4^{2-}. The element was shown to be essential for biological nitrogen fixation in the 1930s (the reduction of N_2 to ammonia) which occurs in many legumes (peas, beans, alfalfa, etc.).

The major source of nitrogen for non-legumous plants is the nitrate ion, NO_3^-. Another molybdenum-containing enzyme, nitrate reductase catalyses the reduction of NO_3^- to NO_2^- in the first step in this pathway for assimilation of inorganic nitrogen.

Molybdoenzymes also play several important roles in animal metabolism. Xanthine oxidase and xanthine dehydrogenase (Table 1.6), catalyse the oxidation of xanthine (**1.5**) to uric acid (**1.6**), a key step in purine metabolism; excessive production of uric acid is one of the primary causes of gout. Other reactions catalysed by molybdoenzymes include the conversion of aldehydes to carboxylic acids by aldehyde oxidase and the oxidation of SO_3^{2-} to SO_4^{2-} by sulphite oxidase (Table 1.6).

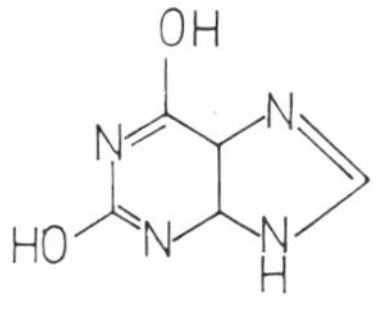

(1.5)

Xanthine

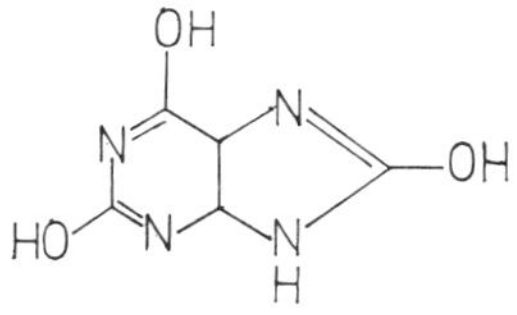

(1.6)

Uric acid

The most highly oxidized molybdenum compounds are formally Mo(VI) species and are e.s.r. silent as the Mo atom has a $4d^0$ configuration. Complexes of Mo(VI) have been studied in detail because (a) Mo(VI) is the form in which the metal is absorbed by plants and animals, and (b) Mo(VI) is thought to be present in the resting state of xanthine oxidase. The tetrahedral molybdate ion (MoO_4^{2-}) will readily form polymers containing the six-coordinate metal ion if the pH is below 7 and the concentration is sufficiently high.

Table 1.6 – Some examples of molybdenum containing enzymes.

Enzyme	Source	Molecular weight	Number of Mo per enzyme	Additional groups in enzyme	Reaction catalyzed
Nitrogenase	*Clostridium*				
Fe-Mo protein	*pasturianum*	220,000	2	24Fe, 24S	$N_2 + 6H^+ + 6e^- \rightarrow 2NH_3$
Fe-protein	(anaerobic bacteria)	56,000	0	4Fe, 4S	
Nitrate reductase	*Neurospora crassa* (fungus)	228,000	1–2	Cytochrome *b* FAD	$NO_3^- + 2H^+ + 2e^- \rightarrow NO_2^- + H_2O$
Xanthine oxidase	Cows milk	275,000	2	8Fe 8S 2 FAD	Xanthine + $H_2O \rightarrow$ uric acid + $2H^+ + 2e^-$
Xanthine dehydrogenase	Chicken liver	300,000	2	8Fe, 8S 2 Flavin	Xanthine + $H_2O \rightarrow$ uric acid + $2H^+ + 2e^-$
Aldehyde oxidase	Rabbit liver	270,000	2	8Fe, 8S 2 FAD	$RCHO + H_2O \rightarrow RCOOH + 2H^+ + 2e^-$
Sulphite oxidase	Bovine liver	110,000	2	2 Heme	$SO_3^- + H_2O \rightarrow SO_4^{2-} + 2H^+ + 2e^-$

The structures of Mo(VI) coordination compounds provide some information about the possible *in vivo* coordination environment. Most monomers, dimers and polymers are six-coordinate and contain at least one oxygen atom multiply bonded to Mo. When more than one terminal oxygen is present the oxygens are always located *cis* to each other as in [$MoO_2(dtc)_2$] (dtc = diethyldithiocarbonate, (**1.7**) and $K_2[Mo_2O_5(C_2O_4)_2(H_2O)_2]$ (**1.8**).

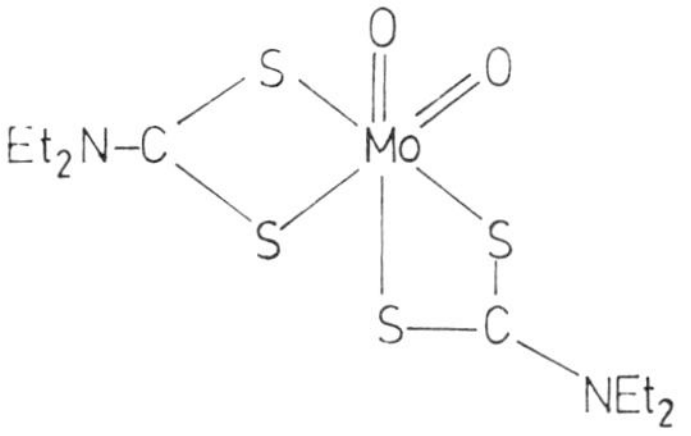

(**1.7**)

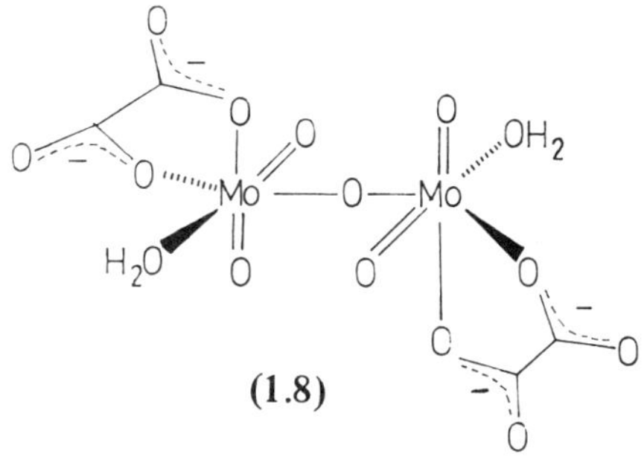

(**1.8**)

Mo(V) has a 4d[1] configuration and should give an e.s.r. signal. Reduction of aqueous solutions of MoO_4^{2-} by thioglycolic acid gives solutions which exhibit e.s.r. signals whose e.s.r. parameters ($\langle g \rangle = 1.978$; $\langle A \rangle = 36G$) are similar to those of xanthine oxidase. Other sulphur-containing ligands also reduce MoO_4^{2-} to give e.s.r. active solutions. Such studies suggest that the molybdenum centre of xanthine oxidase contains a Mo(V) species coordinated by at least one S atom.

A variety of binuclear Mo(V) complexes have been prepared, and some examples are shown in Table 1.7. The main features of the binuclear compounds

Table 1.7 – Examples of binuclear molybdenum(V) compounds

Compound	Structure	Mo . . . Mo Distance (Å)
$Mo_2O_3(S_2CR)_4$	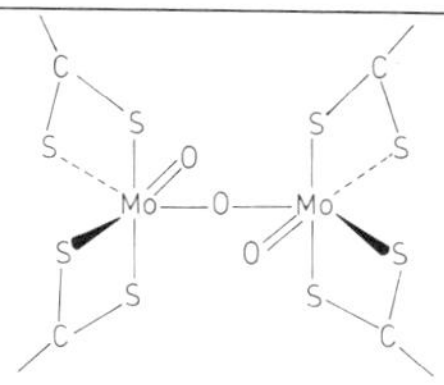	3.86
$[Mo_2O_4(cys)_2]^{2-}$	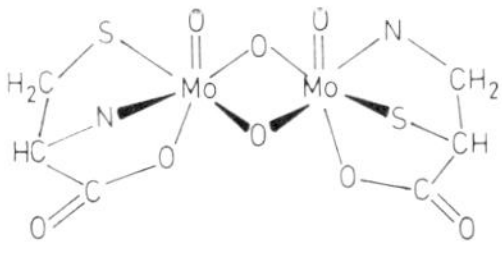	2.57
$Mo_2O_2S_2(his)_2$	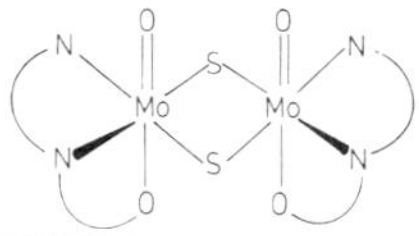	2.82

are: (a) a strongly bound terminal O or S atom attached to each Mo atom; (b) 1, 2 or 3 bridging atoms between the two Mo atoms; and (c) coordination numbers of five, or six with the six-coordinate complexes showing a long Mo–ligand distance for the ligand *trans* to the terminal O or S atom.

All of the binuclear complexes containing bridging O and S atoms are diamagnetic and give no e.s.r. spectrum. As a result, these compounds are not suitable models for the molybdenum centre of xanthine oxidase during turnover. Such binuclear complexes may be models for the resting enzyme.

Manganese

Manganese(II) is found in a variety of enzymes such as pyruvate carboxylase and oxaloacetate decarboxylase, where it functions primarily as a Lewis acid. Manganese-containing superoxide dismutases have been isolated from a variety of organisms. A diamine oxidase which contains manganese is also known.

Pyruvate carboxylase from chicken liver has M = 500,000, and consists of four subunits. The enzyme contains one biotinyl group and one Mn(II) per subunit. The overall catalytic process is:

$$\text{E-biotin} + \text{MgATP}^{2-} + \text{HCO}_3^- \longrightarrow \text{E-biotin-CO}_2 + \text{MgADP}^- + \text{P}_1^{2-}$$

$$\text{E-biotin-CO}_2 + \text{Pyruvate} \longrightarrow \text{E-biotin} + \text{oxaloacetate}$$

(E = enzyme)

Manganese is also required for photosynthetic oxygen evolution. In addition, manganese also appears to play an important role in several metabolic processes such as bone growth, glucose tolerance, reproduction, and development of the inner ear.

The Mn(II) ion is d^5, which in the high spin $t_{2g}^3\ e_g^2$ configuration corresponds to the spherically symmetrical ^6S ground state for the free ion. This configuration does not provide ligand field stabilization energy and Mn(II) therefore forms less thermodynamically stable complexes than the succeeding cations of the first row transition series (Fe(II) to Cu(II)).

The Mn(II) ion is a strong oxidant and in the absence of complexing ligands disproportionates to Mn(II) and MnO_2. Even when complexed, Mn(III) remains a strong oxidant and most of its complexes decompose slowly due to oxidation of the ligand. The Mn(III)-phthalocyanine complex binds O_2 in pyridine solution. The final product being μ-oxo-bis(phthalocyanine)manganese(III).

METAL COMPLEXES

The interaction of a metal ion with a ligand can be expressed by the equilibrium

$$M^{n+} + L \rightleftharpoons ML^{n+}; \quad K = \frac{[ML^{n+}]}{[M^{n+}]\,[L]}$$

where K is the formation constant of the complex. Since the 1940s, coordination chemists have been aware of trends in the formation constants of metal complexes, for example the Irving-Williams series of stability (Fig. 1.2). For a given ligand, the formation constant with a divalent metal ion are in the order $Ba^{2+} < Sr^{2+} < Ca^{2+} < Mg^{2+} < Mn^{2+} < Fe^{2+} < Co^{2+} < Ni^{2+} < Cu^{2+} > Zn^{2+}$.

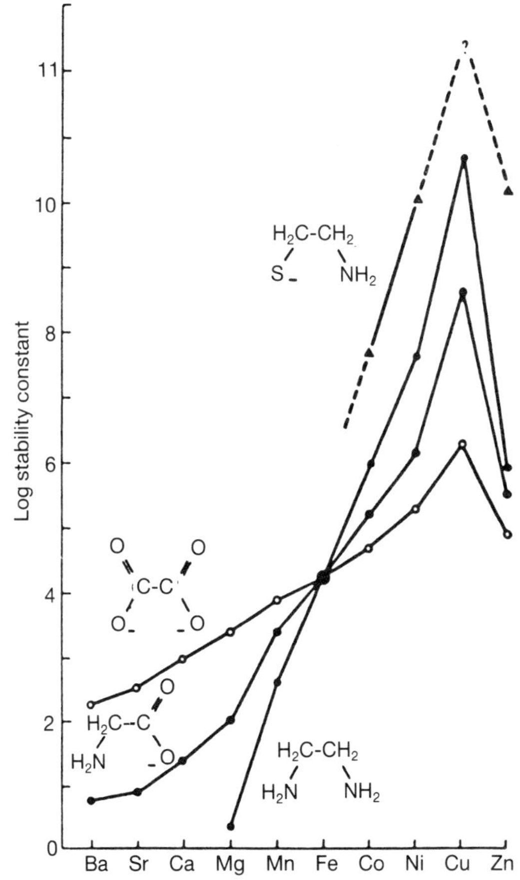

Fig. 1.2 – The Irving-Williams series. The stability increases in the series Ba–Cu, decreases with Zn. [From Sigel, H., and McCormick, D. B., *Acc. Chem. Res.*, **3**, 201 (1970). Reproduced with permission.]

The complexing ability of a cation depends on the ratio of charge to radius and, in the case of transition metal ions, on the ligand field stabilization energy. These factors are illustrated in Fig. 1.3.

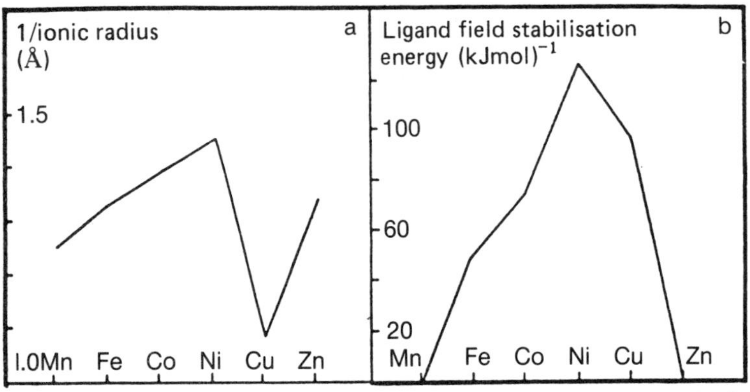

Fig. 1.3 – Factors leading to the Irving-Williams series. From reference [10], reproduced by permission of the Chemical Society, London.

A list of effective radii of metal ions in octahedral coordination is given in Table 1.8.

HARD AND SOFT ACIDS AND BASES (HSAB)

Metal ions are Lewis acids and ligands are Lewis bases. Pearson has developed the concepts of hardness and softness to describe systematically the interaction

Table 1.8 – Effective radii *r* (pm) of metal ions in octahedral coordination†.

	Group I		Group II		Group III		Group IV	
	Li^+	74	Be^{2+}	35	B^{3+}	23		
	Na^+	102	Mg^{2+}	72	Al^{3+}	53	Si^{4+}	26
	K^+	138	Ca^{2+}	100	Sc^{3+}	73	Ti^{4+}	60
	Rb^+	149	Sr^{2+}	116	Y^{3+}	89	Zr^{4+}	72
	Cs^+	188	Ba^{2+}	136	La^{3+}	106	Hf^{4+}	71
	Cu^+	96	Zn^{2+}	74	Ga^{3+}	62	Ge^{4+}	54
	Ag^+	115	Cd^{2+}	95	In^{3+}	79	Sn^{4+}	69
	Au^+	137	Hg^{2+}	102	Tl^{3+}	88	Pb^{4+}	77
			Transition metals					
	Ti^{2+}	86	V^{2+}	79	Ni^{2+}	70		
LS	Cr^{2+}	73	Mn^{2+}	67	Fe^{2+}	61	Co^{2+}	65
HS		82		82		77		73
	Ti^{3+}	67	V^{3+}	64	Cr^{3+}	62		
LS	Mn^{3+}	58	Fe^{3+}	55	Co^{3+}	52	Ni^{3+}	56
HS		65		64		61		60
			Lanthanides					
	La^{3+}	106.1	Ce^{3+}	103.4	Pr^{3+}	101.3	Nd^{3+}	99.5
	Pm^{3+}	97.9	Sm^{3+}	96.4	Eu^{3+}	95.0	Gd^{3+}	93.8
	Tb^{3+}	92.3	Dy^{3+}	90.8	Ho^{3+}	89.4	Er^{3+}	88.1
	Tm^{3+}	86.9	Yb^{3+}	85.8	Lu^{3+}	84.8		

†From Shannon and Prewitt, ***Acta crystallogr., B*** **25**, 925 (1969).

between them. A hard metal ion is one which retains its valence electrons very strongly.

Hard cations are not readily polarized and are of small size and high charge. Correspondingly, a soft cation is relatively large, does not retain its valence electrons firmly, and is easily polarized (Table 1.9).

Table 1.9 – HSAB classification of cations.

Hard	H^+, Li^+ Na^+, K^+, Mg^{2+}, Ca^{2+}, Mn^{2+}, Cr^{3+}, Fe^{3+}, Co^{3+}
Borderline	Zn^{2+}, Cu^{2+}, Ni^{2+}, Fe^{2+}, Co^{2+}, Sn^{2+}, Pb^{2+}
Soft	Cu^+, Ag^+, Au^+, Tl^+, Pd^{2+}, Pt^{2+}, Cd^{2+}

Ligands containing highly electronegative donor atoms (O, N. F) are difficult to polarize and can be classified as hard bases. Such ligands include amines, ammonia, water, and ions such as phosphate or sulphate (Table 1.10). Easily polarized ligands, containing phosphorus, arsenic or sulphur donor atoms, act as soft bases. As a general rule, the formation of stable complexes results from interactions between hard acids and hard bases, or soft acids and soft bases. Hard-soft interactions are weak. The choice of chelating agent for complexing various metals can be rationalized on this basis. The cation K^+ is hard with a noble gas structure and will therefore interact with hard donors such as oxygen. The bonding will be primarily electrostatic. The ligand [18]-crown-6 (**1.9**) has a hole diameter of 2.6-3.2 Å which provides a good fit for K^+ (2.88 Å) leading

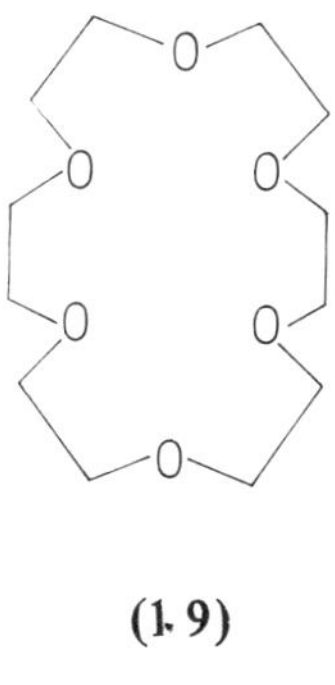

(1.9)

Table 1.10 – HSAB classification of ligands.

Hard	H_2O, OH^-, ROH, OR^-, R_2O, NH_3, NCS^-, Cl^-, PO_4^{3-}, SO_4^{2-}, F^-, NO_3^-, CO_3^{2-}
Borderline	Pyridine, RNH_2, N_2, N_3^-, NO_2^-, Br^-
Soft	RSH, RS^-, R_2S, R_3P, R_3As, CO, CN^-, SCN^-, $S_2O_3^{2-}$, H^-, I^-

to a relatively large formation constant (Fig. 1.4). The ligand diethyldithiocarbamate (**1.10**) contains readily polarized sulphur donors and so is suitable for chelation of soft acids such as Cu(I), Cd(II) and Hg(II).

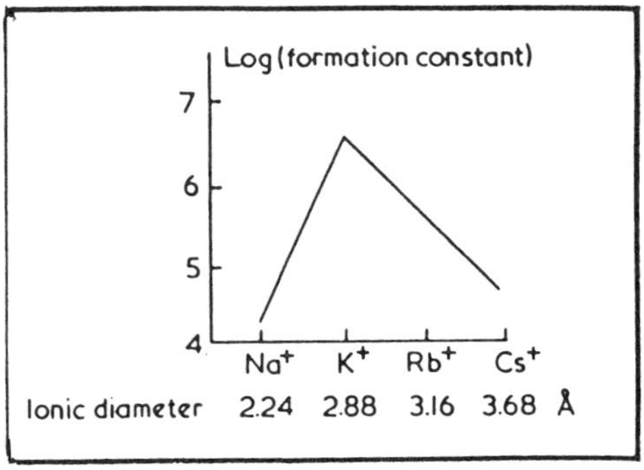

Fig. 1.4 – Formation constants of [18]-crown-6 complexes. From reference [10], reproduced by permission of the Chemical Society, London.

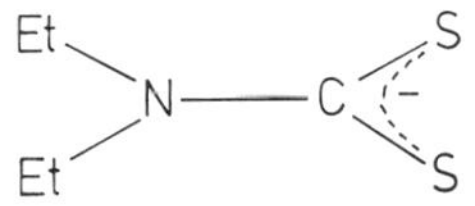

(1.10)

INERT AND LABILE COMPLEXES

Metal complexes can be classified as kinetically inert or kinetically labile. Labile complexes undergo rapid ligand substitution reactions, while inert complexes undergo slow substitution. It should be appreciated that there is no relationship between thermodynamic stability (as determined by a large formation constant) and kinetic inertness. Thus $[Ni(CN)_4]^{2-}$ has a very high formation constant, but the CN^- ligands undergo rapid exchange with $^{14}CN^-$ in aqueous solution.

A very wide span of rates is found, ranging from the extremely slow exchange of NH_3 with $[Co(NH_3)_6]^{3+}$ in aqueous solution (no exchange in 162 days), to the almost diffusion controlled exchange of H_2O between $[Cu(H_2O)_6]^{2+}$ and water ($t_{1/2} \sim 10^{-8}$ s). Substitution reactions dominate in coordination chemistry. Substitution is not infrequently the first step in a redox process, or in a dimerization of polymerization reaction. The process is of importance in the reactions of metal or metal-activated enzymes and the transport of metal ions through cell membranes.

Octahedral and tetrahedral complexes normally react by dissociative type mechanisms with bond breaking in the transition state being of primary importance. With square planar complexes the entering ligand plays an important role, with bond making a feature of the activated complex.

Figure 1.5 shows the rate constants for the exchange of solvent water with a variety of hydrated metal ions in solution

$$[M(H_2O)_x]^{n+} \; + \; X\,H_2\bullet \longrightarrow [M(H_2\bullet)_x]^{n+} \; + \; X\,H_2O$$

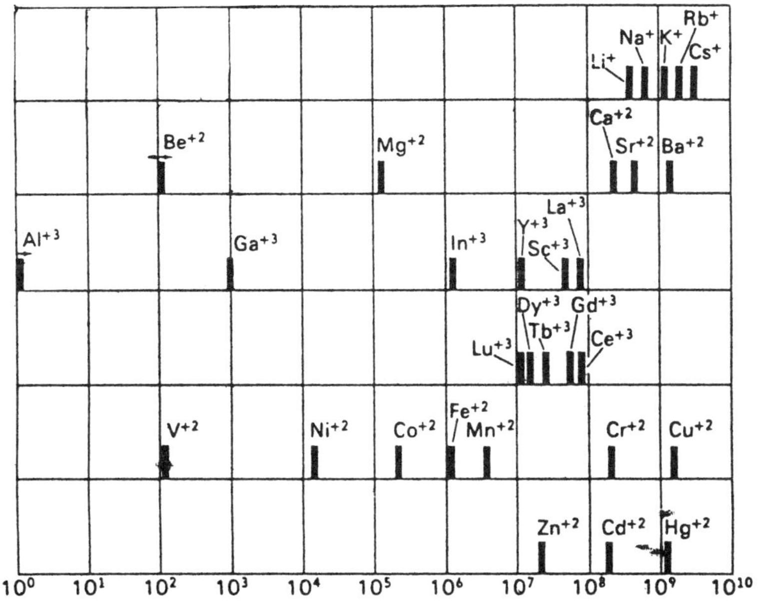

Fig. 1.5 – Characteristic rate constants (sec^{-1}), for substitution of inner sphere H_2O of various aquo ions. [From C. M. Frey and J. Stuehr, "Kinetics of Metal Ion Interactions with Nucleotides and Base Free Phosphates," in "Metal ions in Biological Systems," ed., H. Sigel, Marcel Dekker, Inc., New York, vol. 1, 1974, p. 69. Reproduced with permission.]

For the alkali and alkaline-earth cations where ligand field effects are absent, the rate constants are a function of the ion size for example

	Mg^{2+} <	Ca^{2+} <	Sr^{2+} <	Ba^{2+}
r(pm)	72	100	116	136

	Li^{+} <	Na^{+} <	K^{+}
r(pm)	74	102	138

In octahedral transition-metal complexes the inert metal centres are d^3 (Cr(III)) and low spin d^4, d^5 and d^6 (Co(III)). These complexes have all substantial ligand field stabilization energies. The ligand field stabilization energy is comparable with the activation energies for many of these substitution reactions. The kinetic properties of some aquo metal ions are listed in Table 1.11.

Table 1.11 – Kinetic properties of some metal ions.

V(IV) (d^1)	labile	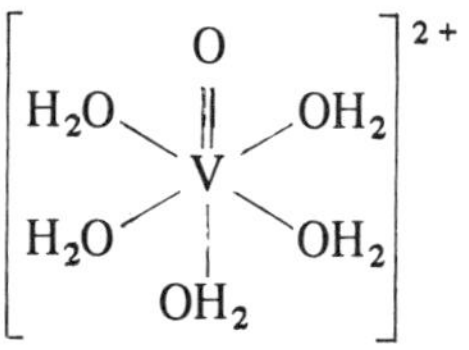
	k (double bonded oxygen)	$\ll 20\ s^{-1}$
	k (equatorial water)	$5 \times 10^2\ s^{-1}$
	k (axial water)	$5 \times 10^8\ s^{-1}$
Cr(III) (d^3)	inert	$[Cr(H_2O)_6]^{3+}$
Co(II) (d^7)	labile	$[Co(H_2O)_6]^{2+}$
Co(III) (d^6, LS)	inert	$[Co(H_2O)_6]^{3+}$
Cu(II) (d^9)*	labile	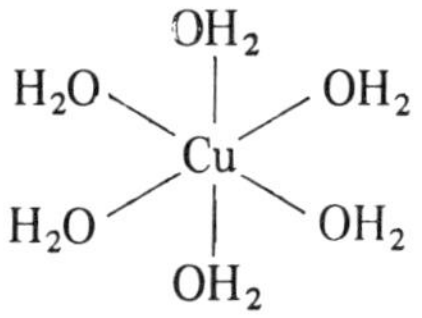

*Rapid exchange of axial water ligands due to Jahn–Teller distortion.

THE STRUCTURE OF PROTEINS

The major constituent of proteins is an unbranched polypeptide chain consisting of S-α-amino acids linked together by amide bonds between the α-carboxyl of one residue and the α-amino group of the next. Normally only the 20 amino acids listed in Table 1.12 are involved in peptide bond formation.

Table 1.12 – The common amino acids.

Amino acid (Three- and one-letter symbols, Molecular weight)	Side chain, R (RCH(NH_3^+)CO_2^-)	pK_as
Glycine (Gly, G. 75)	H–	2.35, 9.78
Alanine (Ala, A, 89)	CH_3–	2.35, 9.87

Valine (Val, V, 117)	CH_3 >CH– CH_3	2.29, 9.74
Leucine (Leu, L, 131)	CH_3 >$CHCH_2$– CH_3	2.33, 9.74
Isoleucine (Ile, I, 131)	CH_3CH_2 >CH– CH_3	2.32, 9.76
Phenylalanine (Phe, F, 165)	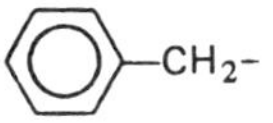	2.16, 9.18
Tyrosine (Tyr, Y, 181)	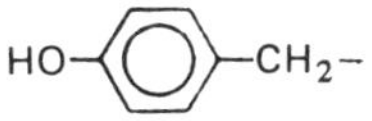	2.20, 9.11, 10.13
Tryptophan (Trp, W, 204)	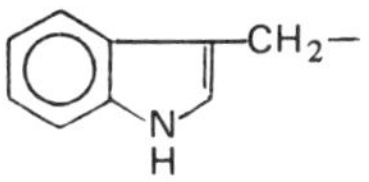	2.43, 9.44
Serine (Ser, S, 105)	$HOCH_2$–	2.19, 9.21
Threonine (Thr, T, 119)	HO >CH– CH_3	2.09, 9.11
Cysteine (Cys, C, 121)	$HSCH_2$–	1.92, 8.35, 10.46
Methionine (Met, M, 149)	$CH_3SCH_2CH_2$–	2.13, 9.28
Asparagine (Asn, N. 132)	$H_2NC(=O)CH_2$–	2.1, 8.84
Glutamine (Gln, Q, 146)	$^-O_2CCH_2$–	1.99, 3.90, 9.90
Aspartic acid (Asp, D, 133)	$^-O_2CCH_2$–	1.99, 3.90, 9.90
Glutamic acid (Glu, E, 147)	$^-O_2CCH_2CH_2$–	2.10, 4.07, 9.47
Lysine (Lys, K, 146)	$H_3N^+(CH_2)_4$–	2.16, 9.18, 10.79
Arginine (Ar, R, 174)	H_2N^+= C–$NH(CH_2)_3$– H_2N/	1.82, 8.99, 12.48
Histidine (His, H, 155)	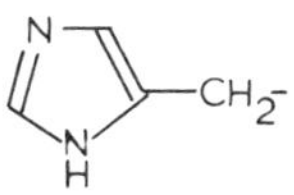	1.80, 6.04, 9.33
Proline (Pro, P, 115)	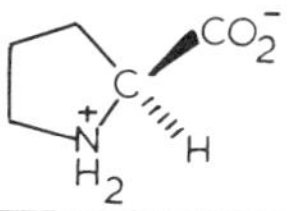	1.95, 10.64

pK_as from *Data for biochemical research,* R. M. C. Dawson, D. C. Elliott, W. H. Elliott, and K. M. Jones, Oxford University Press (1969).

The primary structure is defined by the sequence in which the amino acids form the polymer. By convention, the sequence is written as in (**1.11**), beginning with the N-terminus on the left.

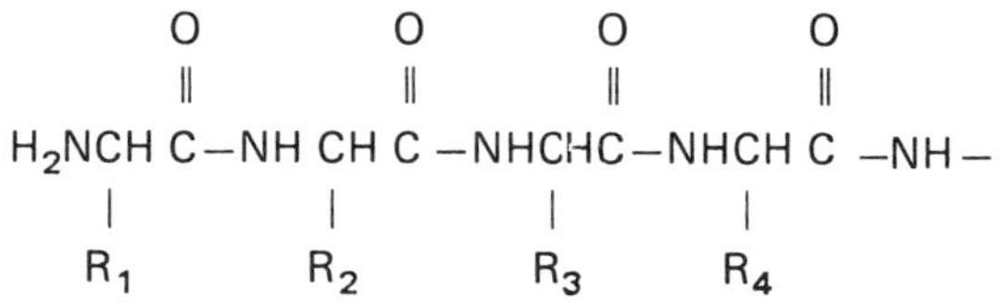

(1.11)

Although the primary structures of almost all intracellular proteins consists of linear polypeptide chains, many extracellular proteins contain —S—S— cross bridges due to oxidative coupling of two cysteine residues ($2RSH - 2H^{+} - 2e \rightarrow R{-}SS{-}R$). Cross-linking may lead to the situations shown in structures (**1.12**) and (**1.13**).

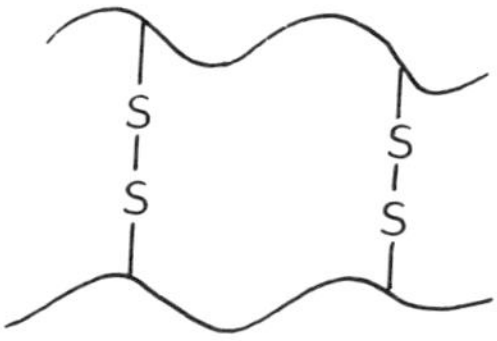

Insulin (1.12)

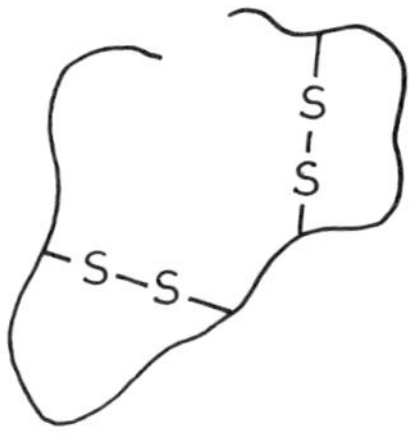

Papain (1.13)

The three-dimensional structure of proteins and enzymes

All enzymes are proteins, but the converse that all proteins are enzymes is of course incorrect. The importance of X-ray diffraction methods in enzymology cannot be overstated. Not only have they provided the experimental basis of our present knowledge of the structure of proteins, but they have proved to be the single most important factor in the investigation of enzyme mechanism.

The degree of accuracy that is attained depends on the quality of the X-ray data and the *resolution*. At low resolution (4~6 Å), the electron density map reveals little more in most cases than the overall shape of the molecule. At 3.5 Å, it is often possible to follow the course of the polypeptide backbone, but there may be ambiguities. At 3.0 Å, it is possible in favourable cases to begin to resolve the amino-acid side chains and, with some uncertainties, to fit the sequence to the electron density. At 2.5 Å, the position of atoms may often be fitted with an accuracy of ±0.4 Å. In order to locate atoms to 0.2 Å, a resolution of about 1.9 Å and very well ordered crystals are necessary. In practical terms this means that the reflections required for high resolution analysis are those that have been diffracted through greater angles and which are of weak intensity. The number of reflections to be analysed increases as the third power of resolution. Thus an increase in resolution from 3 to 1.5 Å increases the amount of data to be collected by a factor of 8. The total effort is increased by an even larger factor due to the poor quality of the data. Table 1.13 summarizes the structural features observable at various degrees of resolution.

Table 1.13 – Resolution and structural information.

Resolution (Å) (1 Å = 0.1 nm)	Structural features observable in a good map
5.5	Overall shape of molecule. Helices as rods of strong intensity.
3.5	The main chain (usually with some ambiguities).
3.0	Start to resolve the side chains.
2.5	Side chains well resolved. The plane of the peptide bond resolved. Atoms located to about ±0.4 Å.
1.5	Atoms located to about ±0.1 Å. The present limit of protein crystallography.
0.77	Bond lengths in small crystals measured to 0.005 Å.

The crystal structures of a number of enzymes, metalloenzymes and proteins have now been determined at 2 to 2.5 Å resolution. Typical examples are ribonucleases (2.0 Å) tetrazinc insulin (2.8 Å), papain, chymotrypsin, and carboxypeptidase.

The peptide bond

X-Ray diffraction studies on the crystals of small peptides have shown that the peptide bond is planar and *trans* (Fig. 1.6). This structure has been found for all peptide bonds in proteins apart from a few rare exceptions. This planarity is due to considerable delocalization of the lone pair of electrons on nitrogen. The C–N bond is consequently shortened and has double-bond character. Twisting of the bond prevents delocalization and some 75 to 88 kJ mol^{-1} of resonance energy is lost. Amides are reasonably acidic, and the pK_a value of the ionization of the peptide hydrogen in the simple dipeptide glycylglycine is estimated to be *ca* 14.5.

$$NH_2CH_2CONHCH_2CO_2^- \rightleftharpoons NH_2CH_2CON^-CH_2CO_2^- + H^+$$

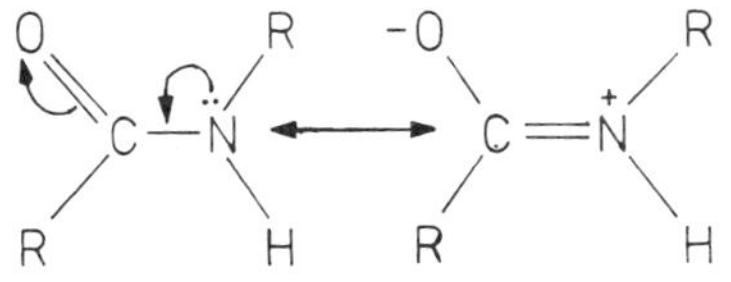

(1.14)

Metal ions may interact with the peptide oxygen or the peptide nitrogen. If a metal ion binds to the peptide nitrogen, the resonance energy of the peptide bond is lost. This resonance energy (*ca* 40 kJ mol^{-1}) can be retained if the peptide

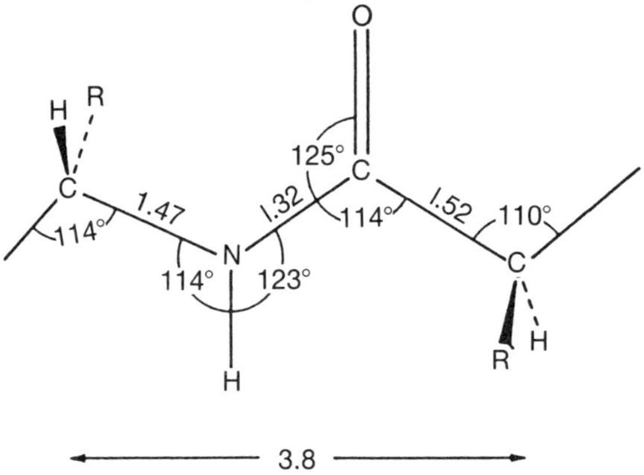

Fig. 1.6 – The peptide bond. Distances are in angstrom units.

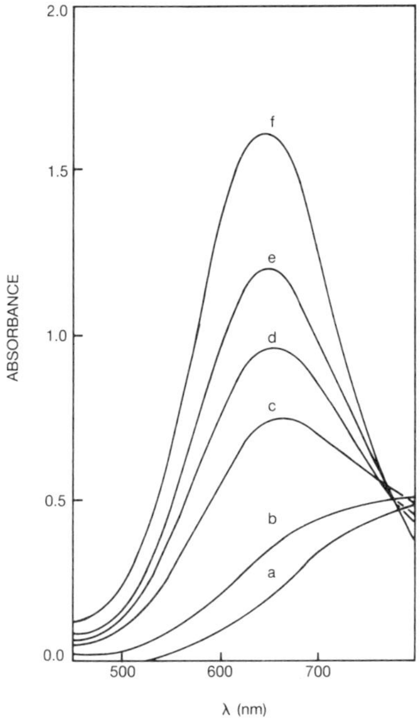

Fig. 1.7 – Visible spectra of Cu(II)-glycylglycine in aqueous solution (a) pH = 3.75; (b) pH = 4.07, (c) pH = 4.52, (d) pH = 4.75, (e) pH = 5,00, (f) pH = 7.76.

hydrogen ionizes, and such deprotonated peptide complexes are observed with many metal ions (Cu^{2+}, Ni^{2+}, Pt^{2+}, Pd^{2+}). This type of complexation can be illustrated by the reaction of copper(II) with glycylglycine. If the pH of a 1:1 mixture of the metal ion and glycylglycine is raised, a new band appears in the visible spectrum at 625 nm (Fig. 1.7) due to the formation of the complex (**1.15**) which has $\epsilon = 84\ dm^3\ mol^{-1}\ cm^{-1}$ at 625 nm.

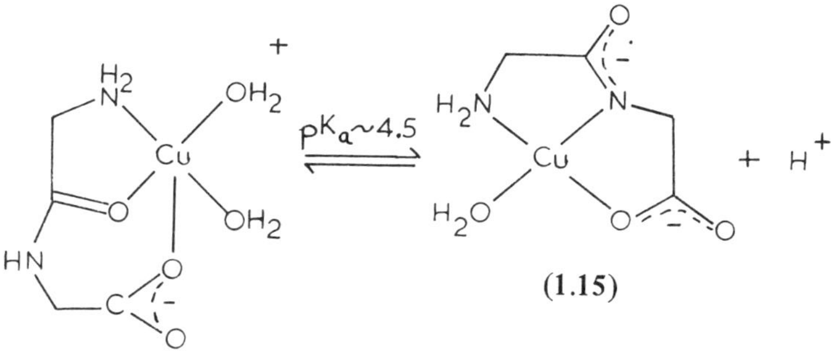

(**1.15**)

Binding Sites in Proteins and Peptides

In addition to the peptide bond itself, the available metal binding sites in proteins are the N-terminal NH_2 group, the C-terminal CO_2^- group and the functional side chains of amino-acid residues such as histidine(His), glutamic acid(Glu), tyrosine(Tyr), cysteine(Cys), aspartic acid(Asp) and of course solvent water. Table 1.14 lists the binding sites of some metalloproteins/enzymes.

Table 1.14 – Metal binding sites[a]

Enzyme/Protein	M^{n+}	1	2	3	4	5	6
Concanavalin A	Mn^{2+}	O(Glu)	O(Asp)	O(Asp)	N(Im)	H_2O	H_2O
Hemerythrin	Fe^{2+}/Fe^{3+}	Four N(Im) and two O(Tyr) are bound to two Fe					
Transferrin	Fe^{3+}	O(Tyr)	O(Tyr)	O(Tyr)	N(Im)	N(Im)	HCO_3^-
Urease	Ni^{2+}		S(Cys)				
Plastocyanin	Cu^{2+}	N(Im)	S(Cys)	N(Im)	N^- (deprot. amide)		
Albumin	Cu^{2+}	NH_2	N^- (amide)	N^- (amide)	N(Im)		
Superoxide dismutase	Cu^{2+}	N(Im)	N(Im)	N(Im)	N(Im)		
Insulin	Zn^{2+}	N(Im)	N(Im)	O	O(H_2O)	O(H_2O)	
Thermolysin	Zn^{2+}	N(Im)	N(Im)	N(Im)	H_2O		
Carboxypepidase A	Zn^{2+}	N(Im)	O(Glu)	N(Im)	H_2O		

[a]Im is the pyridine nitrogen of the imidazole ring of a histidine residue.

The metalloenzyme carboxypeptidase A which contains zinc(II) was the first metalloenzyme in which the ligand binding sites were established (Fig. 1.8).

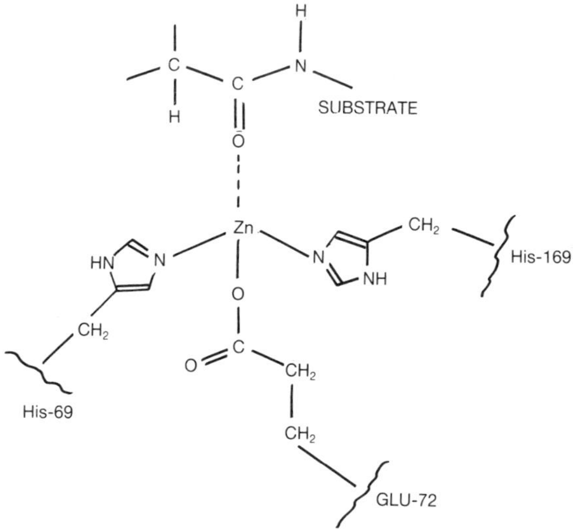

Fig. 1.8 – The coordination of the zinc(II) ion at the active site of carboxypeptidase A.

THE ACTIVE SITE OF ENZYMES

Enzymes are large protein molecules which catalyse organic reactions. The molecular weights of enzymes vary from *ca* 1.2×10^4 to 5×10^5. The substrate molecules are normally small ($M < 10^3$), and so can interact with only a small portion of the enzyme (Fig. 1.9).

X-Ray data on a variety of enzymes indicate that they often have a cleft to which the substrate molecules can bind. This cleft is the active site which contains the catalytic side chains of the amino acids, and other functionalities such as metal ions which are required for catalysis. The enzyme must not bind the products strongly, or the enzyme will be inactivated as the enzyme–product

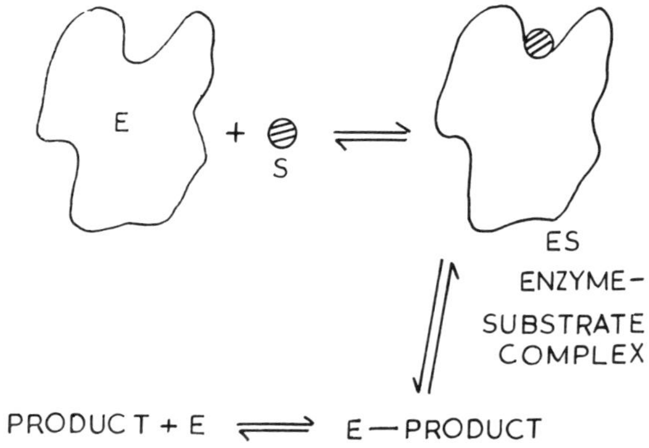

Fig. 1.9 – Schematic diagram of an enzyme-catalysed reaction.

complex (Fig. 1.9). Since the enzyme is a chiral molecule, reaction of the enzyme with only one enantiomer of the substrate is expected, the 'unnatural enantiomer' does not fit into the active site in the required manner.

METALLOENZYMES AND METAL-ACTIVATED ENZYMES

A metal enzyme is one in which the metal ion is an essential participant in the catalytic process. It is convenient to divide metal enzymes into two classes: (1) enzymes with strongly bound metal ions, these are called metalloenzymes and, (2) enzymes with dissociable metal ions, referred to as metal-ion-activated enzymes. Quantitatively the two classes may be distinguished by means of their formation constants. For metalloenzymes the formations constants are $\geqslant 10^8 M^{-1}$, while for the metal-ion-activated enzymes they are $\leqslant 10^8 M^{-1}$. Metalloenzymes can be isolated with the metal ion attached to the protein of the enzyme, whereas the metal ion often dissociates during the purification of the metal-ion-activated enzymes. This leads to a loss of activity, which is usually only regained upon the addition of metal ions of the correct size. In general, the metal ions in metalloenzymes have unusual stereochemistries which most probably contribute to their catalytic behaviour. Some 15 or more cations have been reported to activate various enzymes.

Crystal structure data for various zinc(II) metalloenzymes are listed in Table 1.15. It has been shown that in both the crystalline state and in solution that Zn(II) is in a distorted four-coordinate environment (Fig. 1.8).

NUCLEOSIDES, NUCLEOTIDES AND NUCLEIC ACIDS

Nucleosides are composed of a purine or pyrimidine base attached to the sugar ribose via the N-9 and C-1 atoms respectively. In nucleotides the sugar is linked to a phosphate group, which may be a mono-, di- or tri-phosphate species. The important nucleotide adenosine triphosphate (ATP, Fig. 1.10), illustrates the structural features. This molecule contains the purine base adenine.

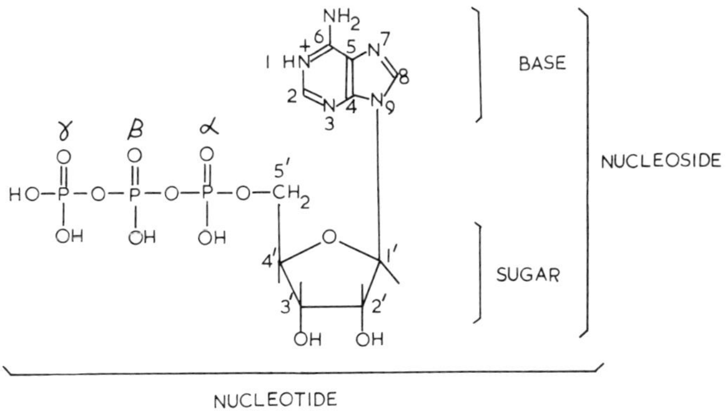

Fig. 1.10 – Nucleotide nomenclature as exemplified for ATP. ATP is shown in the fully protonated form.

Table 1.15 – Crystal structure data for various Zn(II) metallo-enzymes.

Enzyme	Metal Ion(s)	Ligating residues
Alcohol dehydrogenase	Zn	Cys 46, 174
(horse liver)		His 67
	Zn[a]	Cys 97, 100, 103, 111
Carboxypeptidase A	Zn	His 69, 196
		Glu 72
Carbonic anhydrase B	Zn	His 94, 96, 119
Carbonic anhydrase C	Zn	His 93, 95, 117
Superoxide dismutase	Zn	His 61, 69, 78
		Asp 81
	Cu	His 44, 46, 61, 118
Thermolysin	Zn	His 142, 146
		Glu 166

[a]Structural site zincs.

The nucleic acids are polymers built up from nucleotides via phosphodiester bond formation between the 3′-hydroxyl group of one nucleotide and the 5′-hydroxyl group of the adjacent nucleotide. The sequence of the nucleotides is extremely important as it constitutes the genetic code in DNA. Different nucleotides vary in the nature of the purine and pyrimidine base. The four important bases in RNA (ribonucleic acid) are the purines adenine and guanine, and the pyrimidines cytosine and uracil. In deoxyribonucleic acid (DNA) which is a polymer of 2′-deoxyribose nucleotides, thymine(5-methyluracil) is present instead of uracil. The various nucleic acid bases are shown in Fig. 1.11 (R = H in the free bases). The structure of the oxygen containing bases has been written in the lactam form. Lactam ⇌ lactim tautomerization occurs which is dependent on the pH of the environment.

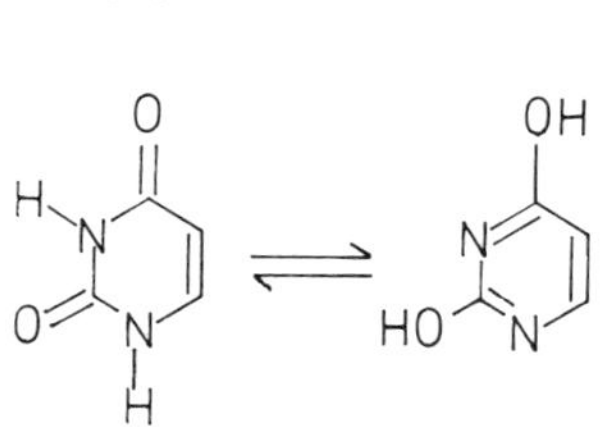

lactam lactim

The structure of RNA is shown diagrammatically in Fig. 1.12, where the different purine and pyrimidine bases are indicated as ‘base’. A greatly simplified means of representing the composition of RNA as Ap Gp Cp Up where A, G, C and U represent the nucleosides adenosine, guanosine, cytidine and uridine respectively and p stands for the phosphate group. The linkage of phosphate to the 3′-hydroxyl group of adenosine is indicated by Ap and the linkage of the phosphate of the 5′-hydroxyl of guanosine is implied by pG.

The biological roles of the nucleic acids and certain nucleotides are dependent on metal ions. Thus Mg^{2+} is involved in the hydrolysis of ATP to ADP + P_i (P_i = inorganic PO_4^{3-}). Base, phosphate and ribose groups are all potential sites for metal ions. Magnesium and calcium ions bind only to the β and γ-phosphate groups of ATP, while Zn(II), Mn(II), Cu(II) and Ni(II) also bind to N-7 of

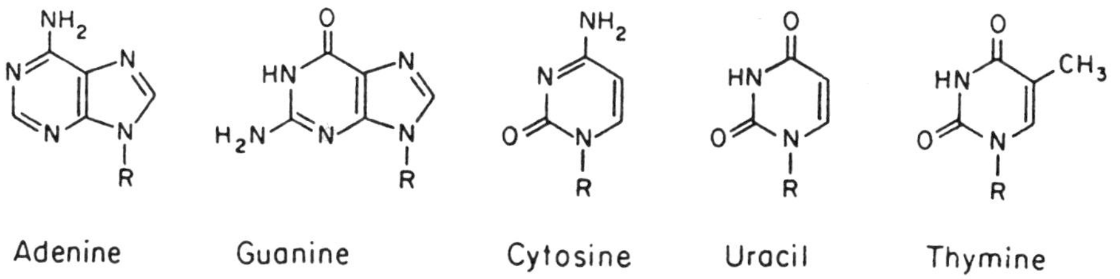

Fig. 1.11 – Nucleic acid bases.

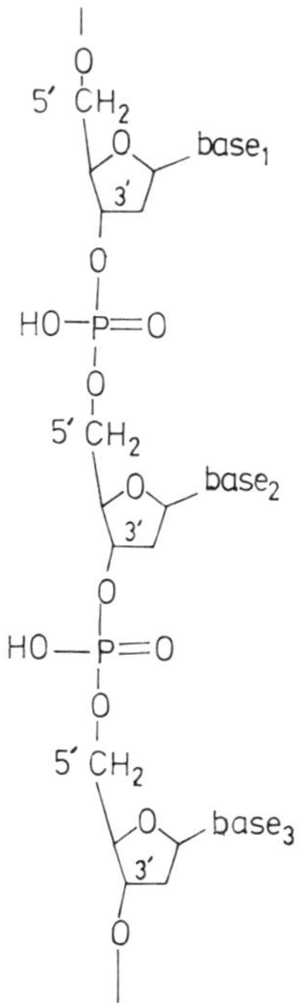

Fig. 1.12 – Structure of RNA.

adenine in addition to phosphate. Hydrolysis of phosphate esters is subject to catalysis by metal ions, thus 8-hydroxyquinoline phosphate is rapidly hydrolysed in the presence of copper(II) ions. Metal ions are widely involved in enzymic phosphate ester hydrolysis. The essential processes of replication, transcription and translation involving the nucleic acids are also dependent on metal ions.

SUGARS

Sugars form complexes with hard cations such as Ca^{2+}. The approximate formation constant of the *epi*-inositol calcium complex which has the structure (**1.17**) is *ca* 3 M^{-1}.

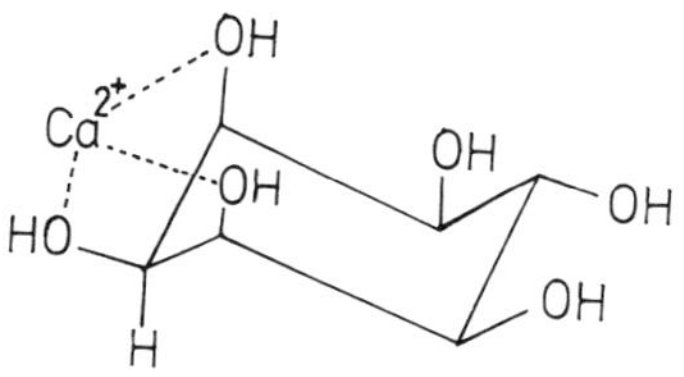

(1.17)

Either electrophoresis or n.m.r. techniques can be used to detect complex formation. (At high pH sugars migrate towards the anode, as they are partially ionized. However, in the presence of Ca^{2+} they migrate to the cathode due to complex formation.)

The sugars which form stable complexes (e.g. allose, talose, ribose) are relatively rare in nature. However, polysaccharides offer the possibility of complexing to more than three oxygen atoms. Alginic acid, is a polysaccharide which forms strong gels in the presence of Ca^{2+}. The molecule is made up long chains of β-D-mannuronic acid and α-L-guluronic acid. The higher the proportion of guluronic acid, the better the gel-forming properties of the alginic acid. A possible structure of the polysaccharide calcium complex is shown in Fig. 1.13.

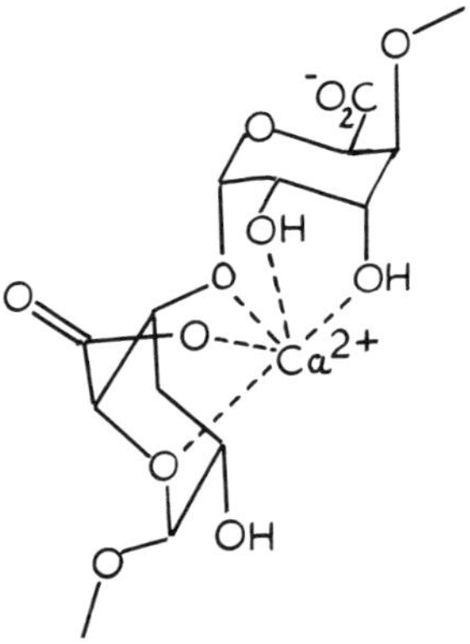

Fig. 1.13 – Possible structure of a section of alginic acid complexed with calcium ion. [Reproduced by permission from *Pure Appl. Chem.*, **35**, 145 (1973)].

BLOOD

Mammalian blood consists essentially of a suspension of living cells (the red and white corpuscles) and particles, known as platelets in an aqueous solution of protein (*ca* 7%), containing metal salts (*ca* 1%), glucose (*ca* 0.1% in humans) and other nutrients with vitamins, hormones and waste products of metabolism in little more than trace amounts. The solution in which the cells and platelets are suspended is known as plasma.

Plasma can be separated from the cells suspended in it by centrifugation, provided precautions are taken to prevent the blood from clotting. Within a few minutes of removal from an animal, blood becomes viscous, and then clots owing to enzymic activity (triggered by Ca^{2+}), which results in the formation of a network of solid fibrous protein (fibrin).

On standing, a clot of blood contracts as the fibrin network shrinks. The clear yellow fluid (plasma deprived of its fibrin) which exudes from the clot is known as serum.

Plasma

The plasma of human blood is a pale, straw coloured, transparent fluid (pH 7.4) containing some 9% of total solids. Plasma has an osmotic pressure equal to that of a 0.9% sodium chloride solution. Such a solution which is isotonic with blood is known as physiological saline.

Plasma Proteins

The plasma proteins account for some 7% of the plasma, and excluding water they are the major constituent. The plasma proteins differ in their isoelectric points and when subjected to electrophoresis, they can be separated into six main fractions (Table 1.16). The albumin, which can be crystallized, accounts for more than 50% of the total plasma protein. As its molecular weight is lower than that of other proteins, it is largely responsible for the osmotic and buffering

Table 1.16 – Plasma proteins.

Fraction	$g/100\ cm^3$ plasma (mean)
Albumin	4.04
α_1-Globulin	0.31
α_2-Globulin	0.48
β-Globulin	0.81
γ-Globulin	0.74
Fibrinogen	0.34
Total	6.72

effects of the plasma proteins. Unlike the albumin and fibrinogen, the globulin fractions are not homogeneous and can be fractionated into many other proteins.

Plasma contains all the amino acids necessary for cell nutrition, glutamine, glutamic acid, alanine and lysine being the most important. The total amino-acid content of human plasma is *ca* 35 mg per 100 cm^3, one-third of this amount being glutamic acid and glutamine. Assuming an average amino-acid molecular weight of ~150, the amino-acid concentration is 2×10^{-3} M. Glycine and alanine comprise about 25% of the total and there are only trace amounts of methionine.

Human plasma may be taken as typical of mammalian plasma, there being no marked differences between different species. A range of inorganic ions are found in blood plasma and their concentrations are summarized in Table 1.1.

BIBLIOGRAPHY

Texts on Bio-inorganic Chemistry

[1] M. N. Hughes, *The Inorganic Chemistry of Biological Process* (2nd edn.), Wiley, London, 1981.

[2] *Inorganic Biochemistry,* ed. G. L. Eichhorn, Elsevier, Amsterdam, 1973 (2 volumes).

[3] H. Sigel, *Metal Ions in Biological Systems,* Vols. 1-20, Marcel Dekker A. G., Basel.

[4] *Techniques and Topics in Bioinorganic Chemistry,* ed. C. A. McAuliffe, Macmillan, London, 1975.

[5] E. J. Underwood, *Trace Elements in Human and Animal Nutrition,* (4th edn), Academic Press, New York, 1977.

[6] *Trace Elements in Human Health and Disease,* ed. A. S. Prasad, Academic Press, New York, 1976.

[7] E. J. Hewitt and T. A. Smith, *Plant Mineral Nutrition,* English Universities Press, London, 1975.

[8] *An Introduction to Bio-inorganic Chemistry,* ed. D. R. Williams, C. C. Thomas, Springfield, Illinois, 1976.

[9] D. A. Phipps, *Metals and Metabolism,* Oxford University Press, 1976.

[10] A. M. Fiabane and D. R. Williams, *The Principles of Bio-inorganic Chemistry,* (Monograph for Teachers, No. 31), The Chemical Society, London, 1977.

[11] P. M. Harrison and R. J. Hoare, *Metals in Biochemistry,* Chapman and Hall, 1980.

[12] E. I. Ochiai, *Bioinorganic Chemistry' An Introduction,* Allyn and Bacon Inc., J. Wiley and Sons, Rockleigh, New Jersey, 1977.

[13] R. P. Hanzlik, *Inorganic Aspects of Biological and Organic Chemistry,* Academic Press, New York, 1976.

[14] *Bioinorganic Chemistry,* ed. R. F. Gould, ACS Advances in Chemistry Series No. 100, 1971. *Bioinorganic Chemistry II,* ed. K. N. Raymond, ACS Advances in Chemistry Series No. 162, 1977.

[15] *Biological Aspects of Inorganic Chemistry,* eds. A. W. Addition, W. R. Cullen, D. Dolphin and B. R. James, John Wiley and Sons, New York, 1977.

[16] A. S. Brill, *Transition Metals in Biochemistry* in *Molecular Biology, Biochemistry and Biophysics,* Vol. 26, ed. A. Kleinzeller, G. F. Springer and H. E. Wittman, Springer-Verlag, Berlin, 1977.

[17] R. J. P. Williams and J. R. R. F. da Silva, *New Trends in Bio-inorganic Chemistry,* Academic Press, London, 1978.

Metal Complexes of Amino Acids Peptides and Proteins

[18] R. W. Hay and D. R. Williams, Specialist Periodical Reports: *Amino Acids Peptides and Proteins,* The Chemical Society, London. (Each odd issue contains a review of the topic.)

[19] H. C. Freeman, Crystal Structures of Metal-Peptide Complexes. *Adv. Protein Chemistry,* **22**, 257 (1967).

[20] R. J. Sundberg and R. B. Martin, 'Interactions of Histidine and other Imidazole Derivatives with Transition Metal Ions in Chemical and Biological Systems', *Chem. Rev.,* **74**, 471 (1974).

Metal Complexes of Sugars

[21] S. J. Angyal, 'Sugar-Cation Complexes – Structure and Applications', *Chem. Soc. Rev.,* **9**, 415 (1980).

Metal Complexes of Nucleic Acids

[22] H. Sigel, *Metal Ions in Biological Systems,* Vol. 8, Marcel Dekker A.G., Basel, 1979.

Metals

Zinc

[23] R. H. Prince, *Adv. Inorg. Radiochem.,* **22**, 349 (1979).

Copper

[24] *Copper Coordination Chemistry, Biochemical and Inorganic Perspectives,* ed. K. D. Karlin and J. Zubieta, Adenine Press, Albany, N.Y., 1983.

[25] C. A. Owen, *Copper Deficiency and Toxicity Aquired and Inherited in Plants, Animals, and Man,* Noyes Publications, Park Ridge, New Jersey, 1981.

Molybdenum

[26] *Molybdenum and Molybdenum Containing Enzymes,* ed. M. Coughlan, Pergamon Press, N. Y., 1980.

[27] K. B. Swedo and J. H. Enemark, 'Some Aspects of the Bioinorganic Chemistry of Molybdenum, *J. Chem. Ed.*, **56**, 70 (1979).

Cobalt

[28] R. W. Hay, 'Cobalt', *Coord. Chem. Rev.*, **35**, 85 (1981); **41**, 191 (1982).

CHAPTER 2

Physical Methods: Illustrative Examples

INTRODUCTION

A variety of physical methods are used to study metalloproteins and metallo-enzymes and a number of illustrative examples are discussed in this chapter. The increasing developments in sophisticated analytical techniques have played an important role in studying metalloproteins. Almost every possible type of analytical method has been applied to the determination of metals in biological systems. Bowen in 1966 listed 78 elements in vertebrate blood, determined by more than ten different procedures, and since then various new techniques have been developed. Neutron activation analysis and the various branches of atomic spectroscopy (atomic emission spectroscopy, atomic absorption spectrophoto-metry) are particularly important, but other techniques such as fluorimetric analysis, electroanalytical methods and electron microprobe analysis all play an important role.

The very great sensitivity of some methods poses a problem in itself. Inadvertent contamination of the material being analysed must be carefully avoided. Sample collection and storage can lead to contamination. For example, blood samples collected by venipuncture using steel needless can add unwanted amounts of chromium and nickel. Many common laboratory materials such as glass, rubber and polyethylene are rich sources of metal contaminants.

Some of the problems of determining metal ion requirements of enzymes can be illustrated by urease, the first enzyme to be crystallized (Sumner 1926). Sumner's work led to the definition of an enzyme as a pure protein with a catalytic function. Only in 1975 was it established by Zerner in Australia that urease is a nickel(II) metalloprotein. Reassessment of the molecular properties of the enzyme shows it to have a molecular weight (M) of 590,000 ± 30,000 which contains six identical subunits (M = 96,600) arranged in the form of an octahedron. Each subunit contains two tightly bound nickel ions (% Ni = 0.12). After the electronic absorption spectrum of the enzyme has been corrected for light scattering, the bands associated with nickel(II) are consistent with nickel in an octahedral coordination environment.

The present chapter will concentrate on a number of spectroscopic techniques which may be used to study metalloproteins.

ELECTRONIC ABSORPTION SPECTRA

In addition to the usual absorption bands present in proteins (at *ca* 280 nm due to the aromatic amino-acid residues L-β-phenylalanine and L-tyrosine), metalloproteins may have additional absorption bands arising from the d–d transitions of the metal ion and charge transfer bands resulting from electronic transitions between the metal ion and the ligand. Thus the nickel(II) metalloenzyme urease has bands at 407 nm, 745 nm and 1060 nm consistent with nickel(II) in an octahedral environment. The d–d transitions in transition in metal ions with unfilled d orbitals are usually of low intensity ($\epsilon \leqslant 1000\ dm^3\ mol^{-1}\ cm^{-1}$) as they are forbidden under the usual selection rules for electronic transitions. The charge transfer bands associated with metal complexes are more intense ($\epsilon > 10^3\ dm^3\ mol^{-1}\ cm^{-1}$) and usually occur in the ultraviolet region.

The absorption spectrum of ascorbic acid oxidase is shown in Fig. 2.1. Two bands are observed, one at 606 nm ($\epsilon = 770\ dm^3 mol^{-1} cm^{-1}$) and a shoulder at

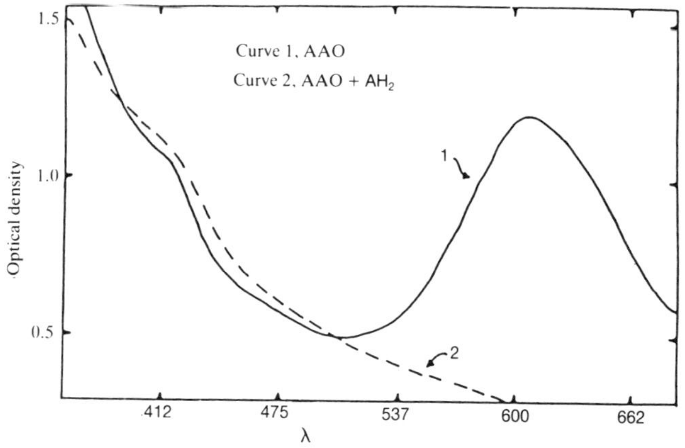

Fig. 2.1 – The visible absorption spectra of ascorbic acid oxidase (AAO), and in the presence of substrate. [From Dawson, C. R., *Ann. NY Acad. Sci.*, **88**, Art 2. 353 (1960). Reproduced with permission.]

412 nm ($\epsilon = 500\ dm^3\ mol^{-1}\ cm^{-1}$). The enzyme obtained from summer crookneck squash (*Cucurbita pepo condensa*) is blue-green in colour and has a molecular weight of 150,000. Analysis indicates 0.26% copper corresponding to six copper(II) atoms per molecule. The enzyme catalyses the oxidation of L-ascorbic acid (**2.1**) to dehydroascorbic acid (**2.2**). When a small amount of ascorbic acid is added to a blue solution of the enzyme in 0.1-M acetate buffer at pH 5.6, the colour is rapidly bleached to a light yellow (Cu(II) → Cu(I), Fig. 2.1). Admission

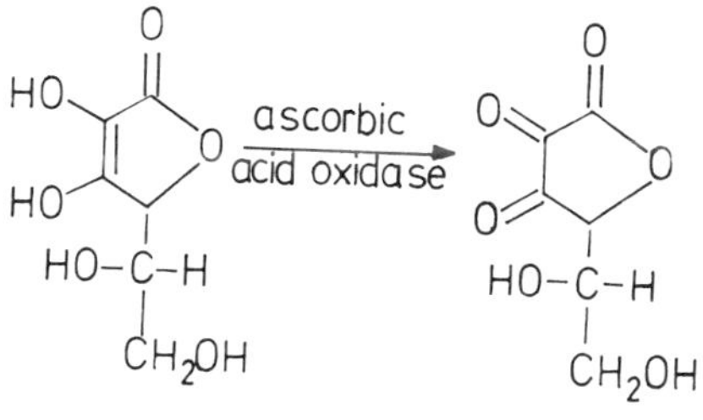

(2.1) (2.2)

of oxygen to the system slowly restores the blue colour due to reoxidation to Cu(II). The absorption spectra of some cobalt(II) and copper(II) metalloenzymes are listed in Table 2.1. Although cobalt(II) metalloproteins are not common, it is possible to substitute cobalt(II) at the active site of some zinc(II) metalloenzymes with the retention of catalytic activity.

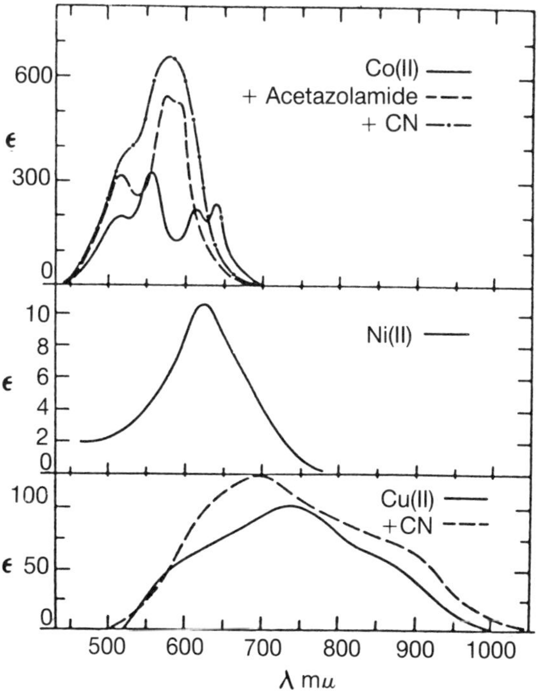

Fig. 2.2 – Visible absorption spectra of Co(II), Ni(II) and Cu(II) human carbonic anhydrases B and inhibitor complexes. [From Coleman, J. E., *Progress in Bioinorganic Chemistry*, Vol. 1. Kaiser, E. T. and Kézdy, F. J. (eds.). Wiley Interscience (1971). Reproduced with permission.] (Note. Acetazolamide is a sulphonamide which binds to the enzyme by coordination at the metal site, a point which has been confirmed by X-ray studies.)

Table 2.1 – Absorption spectra of cobalt(II) and copper(II) metalloenzymes.

Metalloenzymes	Band position, nm		Intensity, ϵ (M^{-1} cm^{-1})
Co(II) carboxypeptidase	500		–
	555		160
	572		160
	940		~25
Co(II) carbonic anhydrase	520		205
	555		340
	615		230
	640		240
	900		~25
Co(II) carbonic anhydrase + CN^-		310[a]	
		345	
		450	
	520	545	350
	570	585	650
Co(II) carbonic anhydrase + acetazolamide		465	
		515	
	520	550	350
	570	570	550
	600	590	500
Co(II) alkaline phosphatase	640		260
	605		220
	555		378
	510		335
Co(II) alkaline phosphatase + HPO_4^{2-}	640		120
	535		350
	480		260
Co(II) alkaline phosphatase + $HAsO_4^{2-}$	500		~240
	550		~260
Cu(II) carboxypeptidase	790		<100
Cu(II) carbonic anhydrase	590		50
	750		100
	900		75
Cu(II) carbonic anhydrase + CN^-	700		130
	900		80
Cu(II) alkaline phosphatase	~750		~100
Copper Oxidases			
Laccase [Cu(II)]	730		~500
	615		1,400
	532		~300
Azurin (psuedomonas blue protein) [Cu(II)]	806		~600
	621		2,800-3,500
	521		~300
	467		~400
Ascorbic acid oxidase [Cu(II)]	606		770
	412		~500
Ceruloplasmin [CuII)]	605		1,200
	370		~500

[a]Figures in this column determined from band positions in the C.D. spectra.

Figure 2.2 shows the visible absorption spectra of cobalt(II), nickel(II) and copper(II) carbonic anhydrases prepared from the native zinc(II) enzyme. Also included are the spectra observed in the presence of some inhibitors of the copper and cobalt enzymes.

The spectrum of the cobalt(II) enzyme is similar to that expected for a tetrahedral complex (ϵ of several hundred), rather than that of an octahedral complex ($\epsilon \sim 10$ dm^3 mol^{-1} cm^{-1}), see Fig. 2.3.

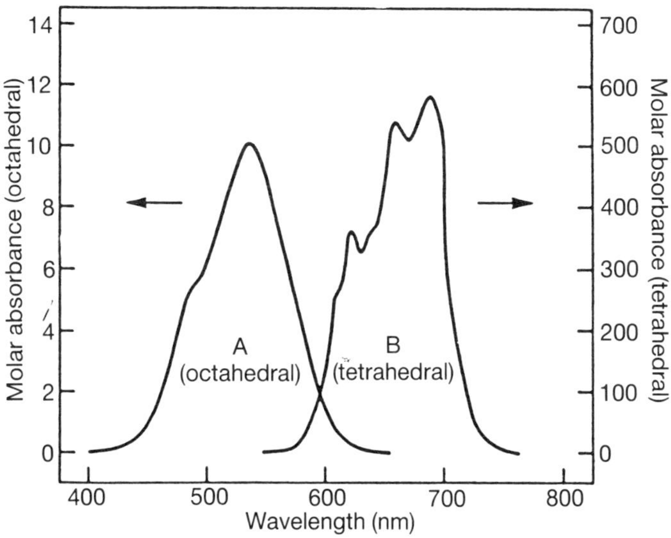

Fig. 2.3 – The visible spectra of $[Co(H_2O)_6]^{2+}$ (curve A) and $[CoCl_4]^{2-}$ (curve B). The molar absorbance scale at the left applied to curve A, and that at the right to curve B. [From Cotton, F. A. and Wilkinson, G., *Advanced Inorganic Chemistry*. Wiley–Interscience. Reproduced with permission.]

The 2 Å resolution X-ray structure of the active site of the enzyme establishes that the zinc enzyme is ligated by the imidazole rings of three histidine residues. The fourth ligand is probably a water molecule or hydroxide ion, depending on the pH. Both the X-ray data and the visible spectrum are consistent with a distorted tetrahedral geometry about the metal ion. The inhibitors cyanide and acetazolamide are both believed to displace the coordinated water molecule. The shift to a narrower more intense spectrum on inhibitor binding has been interpreted as a shift to a more regular tetrahedral geometry.

The visible absorption spectrum of cobalt(II) human carbonic anhydrase B is pH dependent (Fig. 2.4). The change in the spectrum at 640 nm follows a titration curve with pK_a *ca* 7.5, very similar to the pH rate profile of carbonic anhydrase. A good deal of evidence now indicates that this ionization is due to

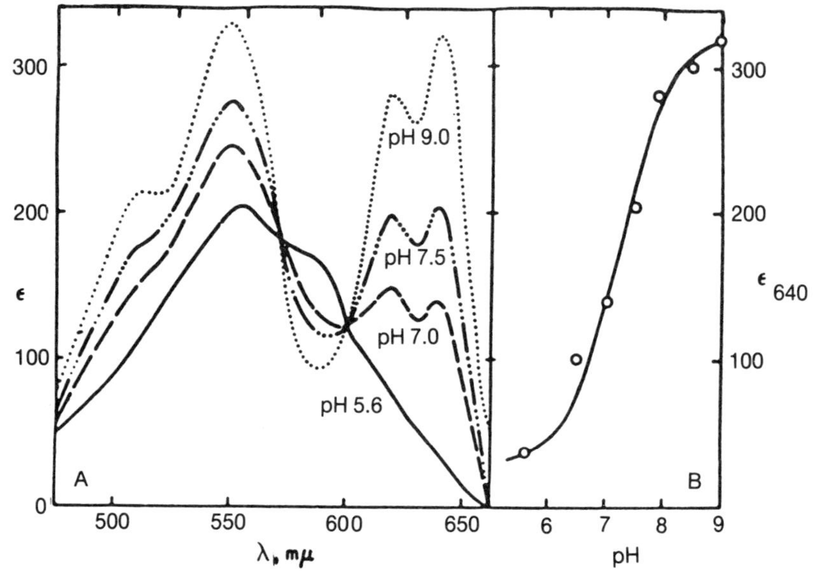

Fig. 2.4 – (A) Visible absorption spectrum of Co(II) human carbonic anhydrase B as a function of pH. (B) Molar extinction coefficient of 640 nm as a function of pH. [From Lindskog, S. and Nyman, P. O., *Biochim. Biophys. Acta,* **85**, 462 (1964)].

$Co{-}OH_2 \rightleftharpoons Co{-}OH + H^+$. A similar ionization is believed to occur in the native zinc(II) enzyme, with hydration of CO_2 occurring by the 'zinc-hydroxide' mechanism (equation 2.1).

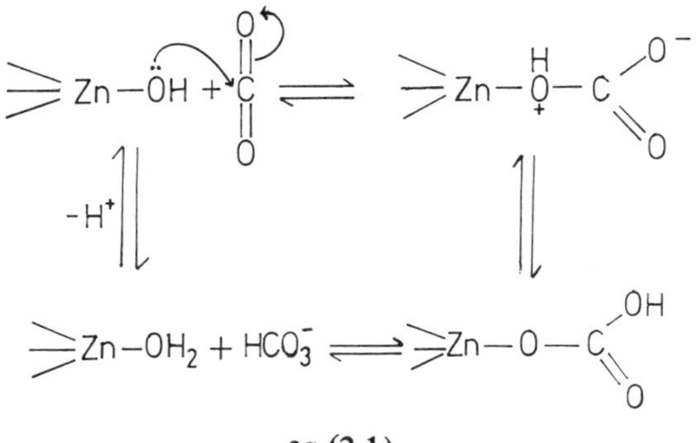

eq (2.1)

OPTICAL ROTATORY DISPERSION AND CIRCULAR DICHROISM

The variation of optical rotation with the wavelength of the plane polarized light is called optical rotatory dispersion (o.r.d). Circular dichroism (c.d.) is defined as the difference in extinction coefficients for left handed and right handed circularly polarized light ($\epsilon_l - \epsilon_r$). A c.d. spectrum is a plot of ($\epsilon_l - \epsilon_r$) versus the wavelength of the circularly polarized light, λ. The combination of absorption, c.d. and o.r.d. is called the Cotton effect (Figs. 2.5 and 2.6). Optical rotatory

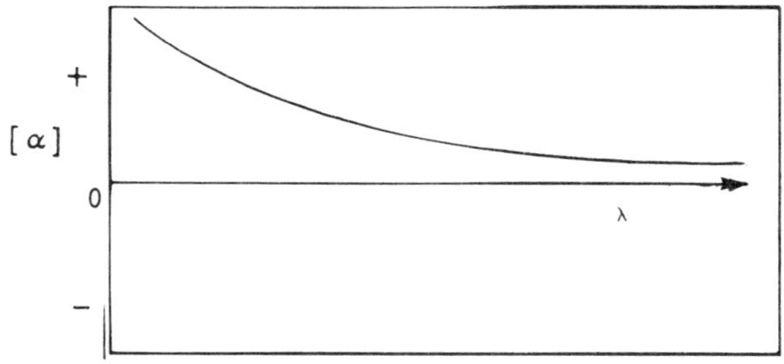

Fig. 2.5 – A plain positive rotatory dispersion curve.

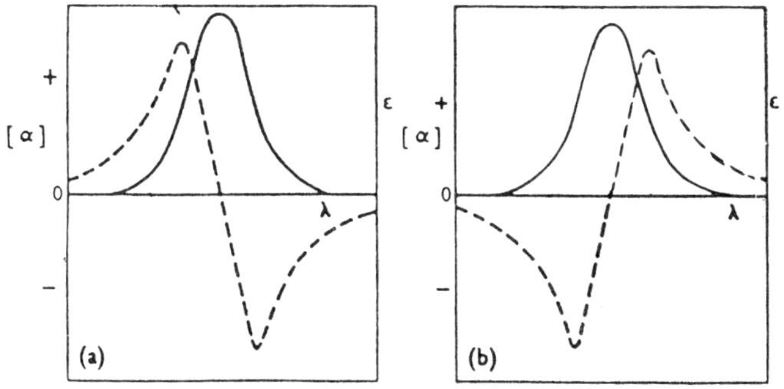

Fig. 2.6 – Rotatory dispersion spectra (------) in the region of an absorption band (——). (a) (–) enantiomer, (b) (+) enantiomer.

dispersion and particularly c.d. measurements have been useful in establishing the absolute configuration of chiral molecules. The absolute configurations of tris-bidentate octahedral complexes are shown in Fig. 2.7.

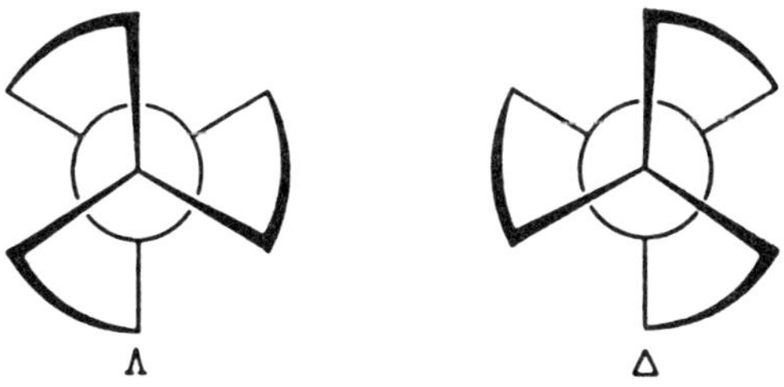

Fig. 2.7 – Absolute configuration Λ and Δ of *tris*-bidentate octahedral species.

The configuration of 1.2.6-(+)[Co(L-ala)$_3$] is shown in Figs. 2.8 and 2.9. This complex has the same handedness as (+)[Co(en)$_3$]$^{3+}$. It should be noted that the (+) sign refers to the sign of the optical rotation at the sodium D line (589 nm = 17,000 cm^{-1}).

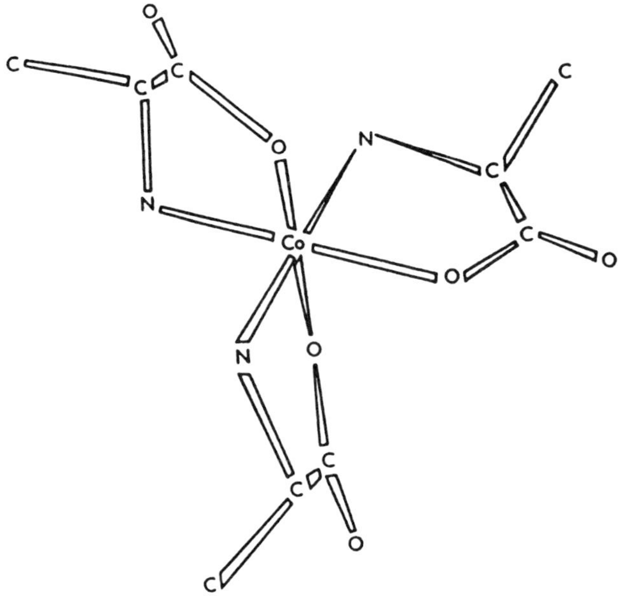

Fig. 2.8 – The absolute configuration of 1,2,6-(+)[Co(L-Ala)$_3$]. [From *Chem. Brit.*, **3**, 205 (1967).

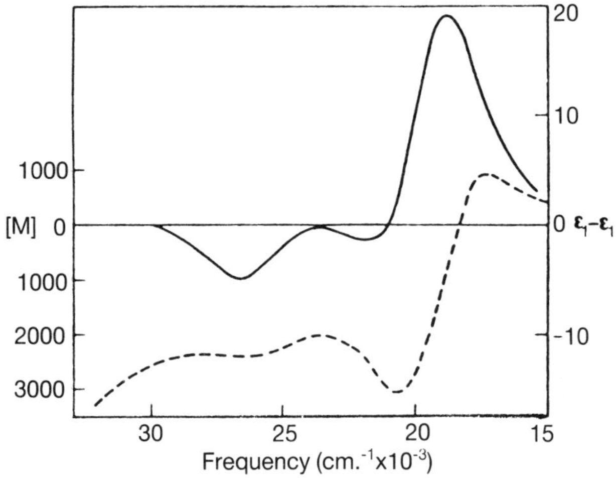

Fig. 2.9 – The Cotton effects of 1,2,6–(+) [Co(L-Ala)$_3$] : ——— circular dichroism and - - - - - optical rotatory dispersion. [From *Chem. Brit.*, **3**, 205 (1967).

Optical rotatory dispersion and circular dichroism measurements have also proved useful in studying metalloenzymes. Circular dichroism in general proves to be much more sensitive to small changes in dissymmetry at the coordination site than the absorption spectrum. Hence, while the the absorption spectra of the cobalt(II) derivatives of human B, human C and bovine carbonic anhydrases are almost identical, only the human C and bovine enzymes show pronounced c.d.'s associated with the d-d bands of cobalt(II) (Fig. 2.10). There are therefore pronounced differences in asymmetry at the active sites of these variants of carbonic anhydrase.

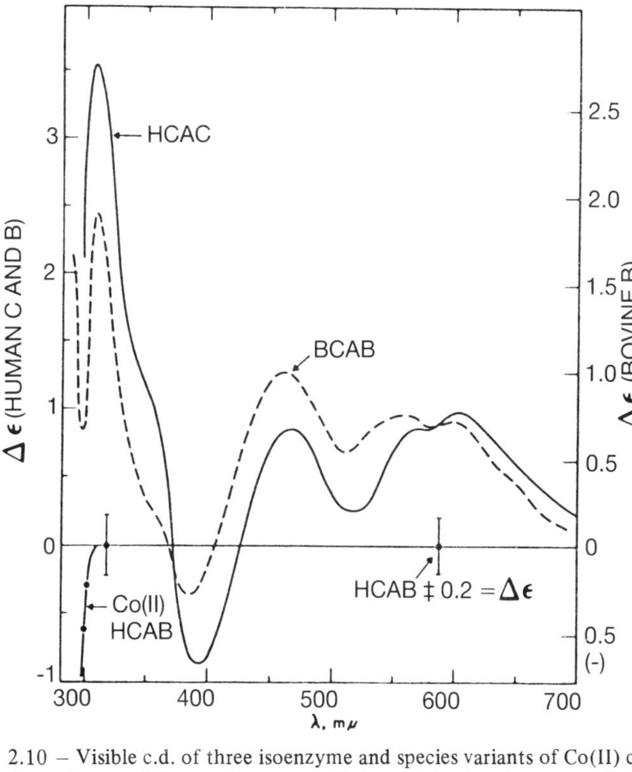

Fig. 2.10 – Visible c.d. of three isoenzyme and species variants of Co(II) carbonic anhydrase. (———) human isoenzyme C: (- - - -) bovine isoenzyme B; (• ——•) human isoenzyme B. [From Coleman, J. E., *Progress in Bioinorganic Chemistry*, Vol. 1. Kaiser, F. T. and Kézdy, F. J. (eds.), Wiley-Interscience (1971). Reproduced with permission].

X-RAY ABSORPTION FINE STRUCTURE (EXAFS)

One of the most powerful and rapidly developing physical techniques for directly probing the metal centres of metalloproteins is X-ray absorption spectroscopy in which synchrotron radiation is used to photoionize core electrons, for example the Mo(1s) electrons in a molybdoenzyme. The fine structure on the absorption peak (EXAFS) contains information about the surrounding atoms encountered by the electron wave as it leaves the metal atom. The method can be used to determine bond lengths, but not bond angles. The general background to the technique can be illustrated using molybdenum compounds. When the energy of the incident X-ray beam exceeds the ionization energy of a 1s electron, this electron is excited to the ionization continuum. As in photoelectron spectroscopy, the energy of the outgoing electron equals the energy of the incoming photon (E_X) less the ionization energy E_0, so that $E_{p.e.} = (E_X - E_0)$. As E_X is continuously increased, the energy of the photoelectron ($E_{p.e.}$) is likewise smoothly increased and its wavelength ($\lambda = hc/E_{p.e.}$, where h = Planck's constant and c is the velocity of light) decreases. The outgoing electron wave, as shown in Fig. 2.11, can undergo back-scattering from atoms surrounding the absorbing atom. The back-scattered electron wave can constructively or destructively interfere with the outgoing wave. This interference gives rise to the EXAFS pattern which is a modulation of the absorption coefficient as a function of the energy of the incident photon. The intensity of the EXAFS depends on the number of scattering atoms at a given distance, and on the magnitude of the scattering power of these atoms.

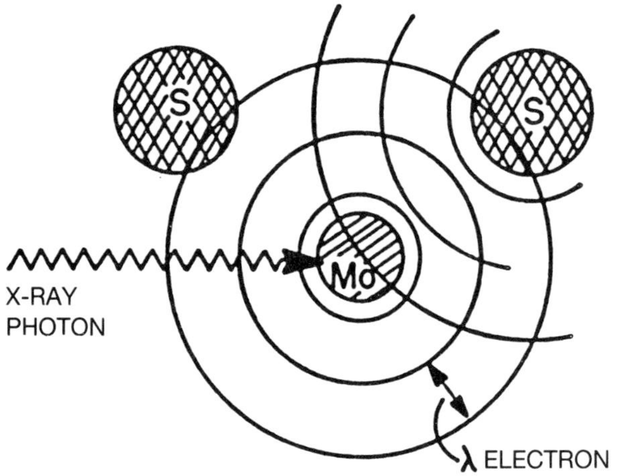

Fig. 2.11 – Diagram illustrating the physical origins of EXAFS. From "Molybdenum and Molybdenum-containing Enzymes", ed. M. P. Coughlan, Pergamon Press, Oxford. Reproduced by permission.

The scattering power is dependent upon the number of electrons in the atom, thus the EXAFS due to scattering from a particular atom (or group of atoms) depends on the nature of the atom(s) and upon the distance of the atom(s) from the absorber. The experimentally observed EXAFS pattern is a superposition of EXAFS due to individual scatters. If the scattering atoms are all of the same type and are at the same distance from the absorber (i.e. a single shell system) then the pattern is relatively simple and its intensity is related to the number of scattering atoms. The presence of two or more atom types or distances causes the overall pattern to contain 'beats' as the individual EXAFS contributions cancel or reinforce.

Fourier transforms of the EXAFS give peaks for the contribution of each shell to the EXAFS. The positions of the peaks on the abscissa of the Fourier transform are related to the scatterer-absorber distance, and the amplitude of the peaks to the scattering power of the shell, i.e. to the electron density in the shell. The Fourier transform approximates a (spherically averaged) radial distribution function of electron density around the absorbing atom.

Fourier transforms of single shell systems, i.e. simple Mo complexes in which all donor atoms are equivalent or at the same distance such as $[MoO_4]^{2-}$, give a single strong peak above the noise due to the four equidistant oxygen donors (Fig. 2.12). The transform for $[MoS_4]^{2-}$ is similar except that here the peak is more intense (corresponding to the greater scattering power of S compared with O) and is found at a longer scatterer-absorber distance corresponding to the longer Mo-S versus M-O bond length (2.18 versus 1.76 Å). In the second pair of complexes the intensity and distance each successively increase in agreement with the larger number of scatterers and the increase in metal-ligand distance as one moves from six- to eight-coordinate complexes.

Fourier transforms for the multishell systems $Mo(NHSC_6H_4)_3$, $MoO(S_2CNEt_2)_2$, $MoO_2[CH_3SCH_2CH_2N(CH_2CH_2S)_2]$, $MoO_2[(CH_3)_2NCH_2CH_2N(CH_2CH_2S)_2]$ and $MoO_2(NH_2CH_2CH_2NHCH_2CH_2NH_2)$ are shown in Fig. 2.13. The second complex clearly shows distinct Mo-O and Mo-S shells, but the remaining complexes show decreased resolution of individual shells and it becomes difficult to assign individual shells to particular scattering groups. In order to get accurate structural data from multishell systems various curve fitting procedures have been developed. The curve fitting procedure devised at Stanford University by Hodgson, Cramer and their coworkers [7] has been used to study Mo complexes and enzymes. The curve fitting procedures provide an approximate value for the number of scattering atoms and a precise value of their distances. Results for some Mo complexes are listed in Table 2.2, where it can be seen that EXAFS can determine Mo-X distances to within 0.02 Å or better. The number of scattering atoms at the accurately determined distance is only specified to within *ca* 20%.

If EXAFS data on metalloproteins are to be accepted as providing precise data on the metal coordination sphere it should be possible to determine bond distances accurately in complexes of unknown crystal structure. This type of work has been carried out. For the complex $MoO[CH_3SCH_2CH_2N(CH_2CH_2S)_2]$ the EXAFS data are given in Table 2.3. The crystal structure shown in Fig. 2.14 was subsequently determined. The close agreement gives confidence in EXAFS distance determinations for the first coordination sphere of Mo in enzymes.

Table 2.2 – EXAFS Analysis[a] of complexes and known structures.

Complex	Mo-O[b]		Mo-S[c]		Mo-N[d]		Mo-Mo[e]	
	X-ray	EXAFS	X-ray	EXAFS	X-ray	EXAFS	X-ray	EXAFS
MoS_4^{-2}			2.18(4)	2.18(4.7)				
Mo $(S_2CN(C_2H_5)_2)_4$			2.53(8)	2.53(6.6)				
MoO_3 $(NH_2CH_2CH_2NHCH_2CH_2NH_2)$	1.74(3)	1.74(3.3)			2.33(3)	2.33(2.2)		
MoO $(S_2CNR_2)_2^f$	1.66(1)	1.66(1.2)	2.41(4)	2.43(3.7)				
$Mo_2O_4(S_2CN(C_2H_5)_2)_2^{-2}$	1.68(2)	1.67(1.3)	2.45(2)	2.45(1.5)			2.58(1)	2.58(1.0)

a Data from Cramer *et al.* (1978). Numbers in parenthesis are the number of atoms, *N*, found at the given distance, *b* MoO_4^{-2} used as calibrant, *c* $MO(S_2C_6H_4)_3$ used as calibrant, *d* $Mo(NCS)_6^{-3}$ used as calibrant, *e* Mo_2O_4 $(cysteinate)_2^{-2}$ used as calibrant, *f* $R = C_2H_5$ for EXAFS but $R = n\text{-}C_3H_7$ for the crystallography.

Table 2.3 – Comparison of Distances in Å determined by EXAFS and X-ray crystallography in $MoO_2[CH_3SCH_2CH_2N(CH_2CH_2S)_2]$.

Bond	EXAFS	X-ray
Mo-O	1.693(2.1)	1.694(2)
Mo-S	2.401(2.7)	2.405(2)
Mo-S	2.803(0.5)	2.809(1)

The EXAFS data were reported prior to the X-ray data. Numbers in parenthesis refer to the number of atoms found at the given distance.

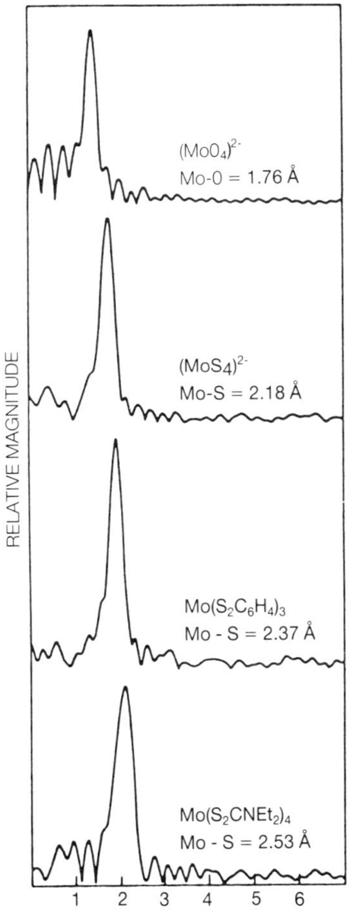

Fig. 2.12 – Fourier transforms for single shell systems. From "Molybdenum and Molybdenum-containing Enzymes", ed. M. P. Coughlan, Pergamon Press, Oxford. Reproduced by permission.

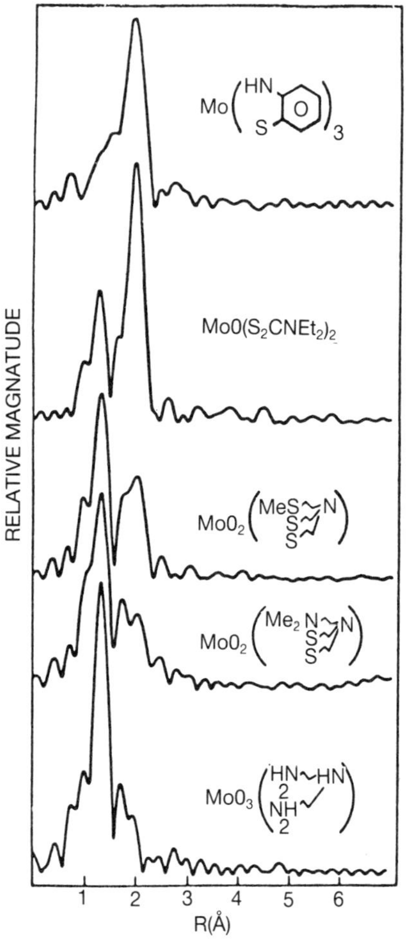

Fig. 2.13 – Fourier transforms for multishell systems. From "Molybdenum and Molybdenum-containing Enzymes", ed. M. P. Coughlan, Pergamon, Press, Oxford. Reproduced by permission.

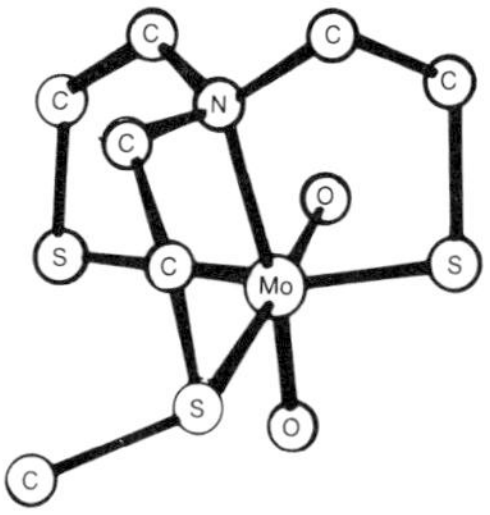

Fig. 2.14 – The molecular structure of $MoO_2[CH_3SCH_2CH_2N(CH_2CH_2S)_2]$ (Berg, *et al.*, 1979).

The ability of EXAFS to distinguish details of the coordination sphere of Mo is illustrated by direct comparison of the EXAFS pattern of $MoO_2[(CH_3)_2NCH_2CH_2N(CH_2CH_2S)_2]$, $MoO_2[CH_3SCH_2CH_2N(CH_2CH_2S)_2]$, sulphite oxidase [Mo(VI) form] and xanthine oxidase [Mo(VI) form] shown in Fig. 2.15. The first two EXAFS curves (they are not Fourier transforms) reveals that a difference of a *single* donor atom in the Mo(VI) six-coordination sphere leads to a clear difference in the spectrum and demonstrates the ability of EXAFS to distinguish such small changes. The two bottom curves reveal the difference between the oxidized forms of xanthine and sulphite oxidases. Comparison of the middle two curves shows the great similarity of sulphite oxidase (but not xanthine oxidase) to one of the two very similar model structures.

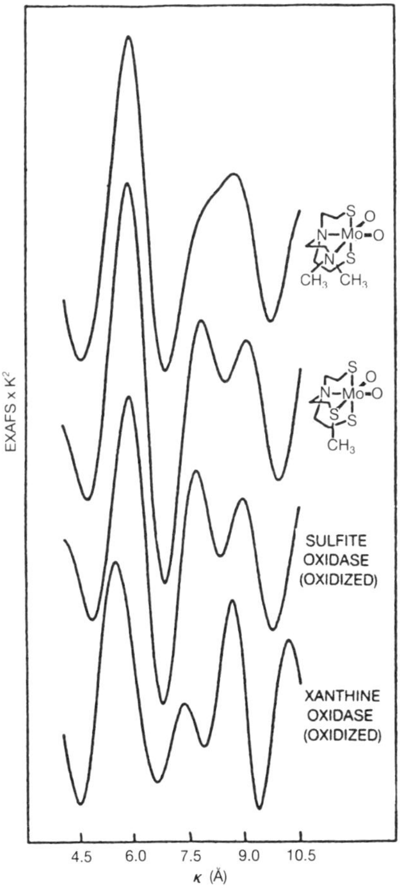

Fig. 2.15 – Comparison of the EXAFS patterns for $MoO_2[(CH_3)_2$-$NCH_2CH_2(CH_2CH_2S)_2]$, $MoO_2[CH_3SCH_2CH_2N(CH_2CH_2S)_2]$, sulphite oxidase [Mo(VI)] and xanthine oxidase [Mo(VI)]. From "Molybdenum and Molbdenum-containing Enzymes", ed. M. P. Coughlan, Pergamon Press, Oxford. Reproduced by permission.

ELECTRON SPIN RESONANCE

Much information has been obtained from e.s.r. measurements. Copper(II) complexes (d^9) with a single unpaired electron, give an easily resolved e.s.r. signal at g-values only slightly higher than that of free electron (g = 2.0023). As copper(I) is d^{10} with no unpaired electrons, e.s.r. measurements can establish that copper(II) is reduced to copper(I) when a substrate is added. Thus the blue copper oxidase, mushroom laccase is reduced on the addition of the substrate catechol (Fig. 2.16).

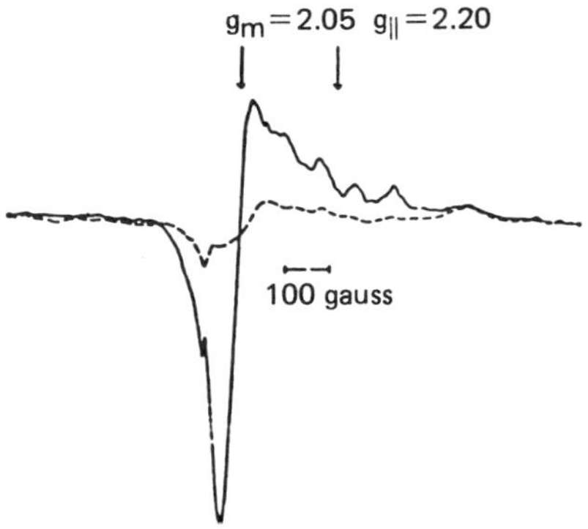

Fig. 2.16 – Electron spin resonance signal of mushroom laccase (———) before and (-----) after treatment with 10 mM catechol. Temperature 77K, field modulation 6.6 gauss. [From Nakamura, T. and Ogura, Y., in *Magnetic Resonance in Biological Systems,* Ehrenberg, A. Malmström, B. G. and Vanngard, T. (eds.), New York, Pergammon Press (1967)].

There are two major nuclear hyperfine interactions which have proved useful in studies of copper(II) binding. The copper(II) signal is split into four lines by the copper nuclear hyperfine interaction (I = 3/2), and if nitrogen is one of the ligand donor atoms, hyperfine lines appear from the $^{14}_{7}N$ nuclei (I = 1). The number of nitrogen hyperfine lines observed is therefore a function of the number of nitrogen donors. The room temperature e.s.r. spectrum of Cu(II) triglycine (**2.3**) is shown in Fig. 2.17. In this complex there are three nitrogen

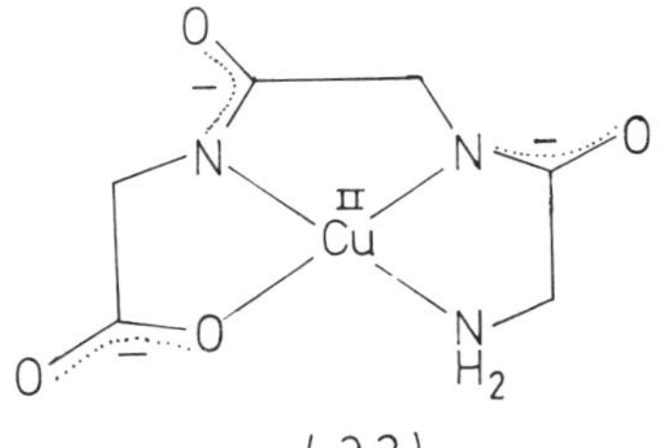

(2·3)

donors and one oxygen donor. Seven hyperfine lines are observed on the high field side of the spectrum, as a result of the three nitrogen donors in addition to the copper hyperfine splitting. (Each e.s.r. signal of an electronic system

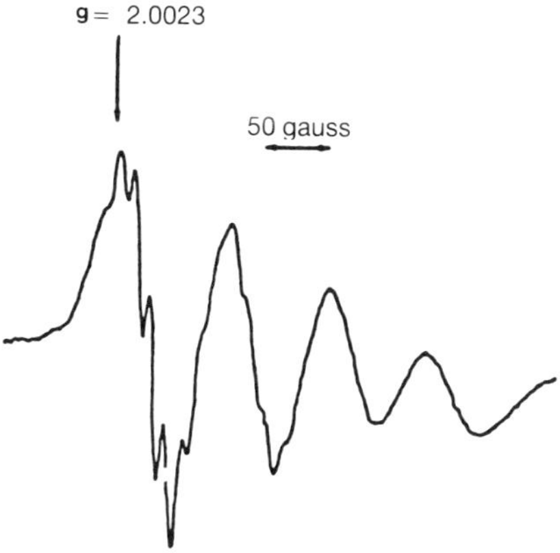

Fig. 2.17 – Room temperature spectrum of [Cu(II) (glycylglycylglycine)]. [From Wiersema, A. K. and Windle, J. J., *J. Phys. Chem.*, **86**, 2316 (1964)].

which interacts with a group of n equivalent nuclei of spin I is split into $(2nI + 1)$ lines, in this case $n = 3$ and for $^{14}_{7}N$, $I = 1$, so that $(2nI + 1) = 7$.) Nitrogen hyperfine splittings have been observed in a number of copper(II) complexes, such as the copper(II)-conalbumines, copper(II)-transferrins, copper(II)-carboxypeptidase and copper(II)-insulin.

The presence of molybdenum in bovine liver sulphite oxidase was discovered accidently as a result of a casual examination of a partially purified enzyme by e.s.r. spectroscopy. The oxidized enzyme does not show any e.s.r. signal at – 180°C, however the enzyme reduced with sulphite has a strong signal at $g = 1.97$ (Fig. 2.18) quite similar to the Mo(V) signals seen in xanthine oxidase

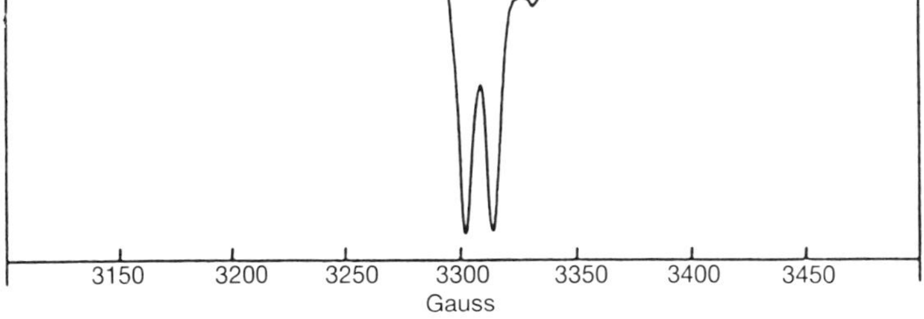

Fig. 2.18 – E.p.r. spectrum of sulphite-reduced bovine liver sulphite oxidase in 0.1 M-Tris-Cl buffer, pH 7.0. From "Molybdenum and Molybdenum-containing Enzymes", ed. M. P. Coughlan, Pergamon Press, Oxford. Reproduced by permission.

and aldehyde oxidase. The presence of molybdenum in the enzyme was confirmed by colorimetric analysis which gave a value of 2 atoms of Mo per molecule of enzyme.

MOSSBAUER SPECTROSCOPY

Mössbauer spectroscopy has been widely used in studies of iron compounds such as hemeproteins and the iron-sulphur proteins. Transitions are observed between two nuclear energy levels of the ^{57}Fe nucleus (a less abundant isotope of iron). These levels are $I = 1/2$ and $I = 3/2$, referring to the magnetic moment of the nuclear spin. Transitions can be observed between these nuclear energy levels corresponding to absorption of γ-radiation from an appropriate γ-ray source (Fig. 2.19).

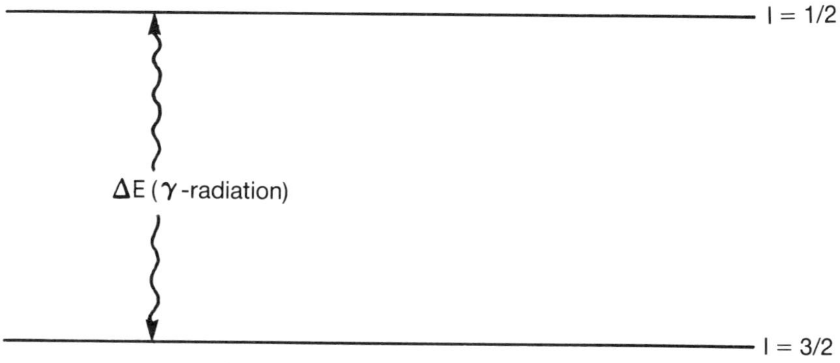

Fig. 2.19 – Nuclear energy levels and absorption of γ-radiation in Mössbauer spectroscopy.

The energy separation, ΔE, depends on both the electronic charge density and the asymmetry of charge at the nucleus. These parameters in turn depend on the electronic state of the iron, i.e. the occupancy of the d orbitals. Characteristic spectra are obtained for different electronic and spin states of iron.

NUCLEAR MAGNETIC RESONANCE

Great advances have been made in the application of high field n.m.r. measurements to proteins and enzymes, and by the use of n.m.r. shift reagents. The use of Fourier transform n.m.r. spectrometers operating at 270 MHz or greater, coupled with a variety of new difference spectroscopy techniques has made it possible to carry out procedures normally used in the examination of the n.m.r. spectra of small molecules.

Proteins which have now been studied in some considerable detail by such high resolution n.m.r. techniques include lysozyme, cytochrome c, cytochrome c_3, triose phosphate isomerase, carbonic anhydrase, peroxidase and some copper and calcium proteins. The lysozyme molecule is shown in Fig. 2.20, and a portion of the ^{1}H-n.m.r. spectrum in Fig. 2.21, with the various assigned amino-acid residues.

The inspection of comparative structural features by ^{1}H-n.m.r. is very fast, as the n.m.r. spectrum provides a structural fingerprint of the molecule. For example, the active site of human leukaemia lysozyme was shown to be closely related to that of hen egg-white lysozyme in work which took a single day.

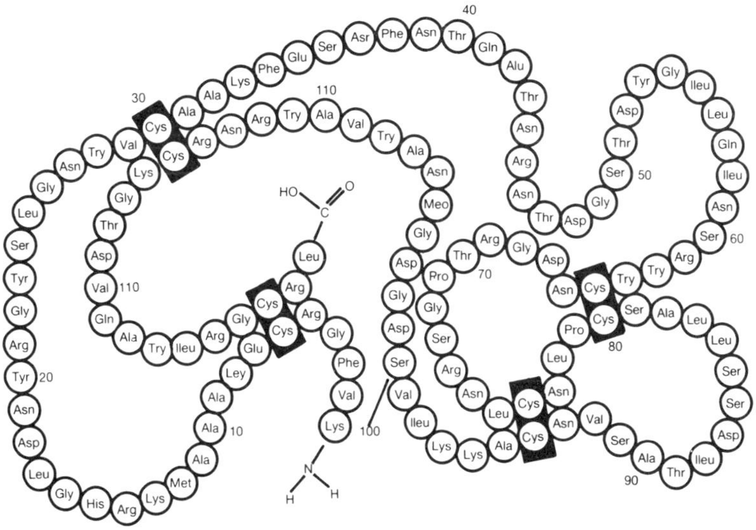

Fig. 2.20 – Amino-acid sequence of the enzyme lysozyme as determined by P. Jollès (*Proc. R. Soc.*, **167B**, 350 (1967); R. E. Canfield, *J. Biol. Chem.*, **238**, 2698 (1963). By courtesy of R. E. Canfield and the American Society of Biological Chemists.

On the basis of a definitive assignment of many individual proton signals in the n.m.r. spectrum of lysozyme (Fig. 2.21), and the use of inorganic relaxation probes [Mn(II), Gd(III), $[Cr(CN)_6]^{3-}$], and shift probes [Pr(III), Co(Porphyrins), $[Fe(CN)_6]^{3-}$] it has been possible to establish that the main chain fold of lysozyme in solution is similar to that in the solid state.

X-RAY CRYSTAL STRUCTURE DETERMINATION

X-Ray crystallography is the most powerful technique for studying crystalline proteins. Much of our knowledge of the active site of metalloenzymes and the structure of metalloproteins comes from work of this type. Some of the problems associated with the technique are outlined in Chapter 1. The number of completed structures of relatively small enzymes (M = 30,000) in crystalline form is about 30 at the present time (1983).

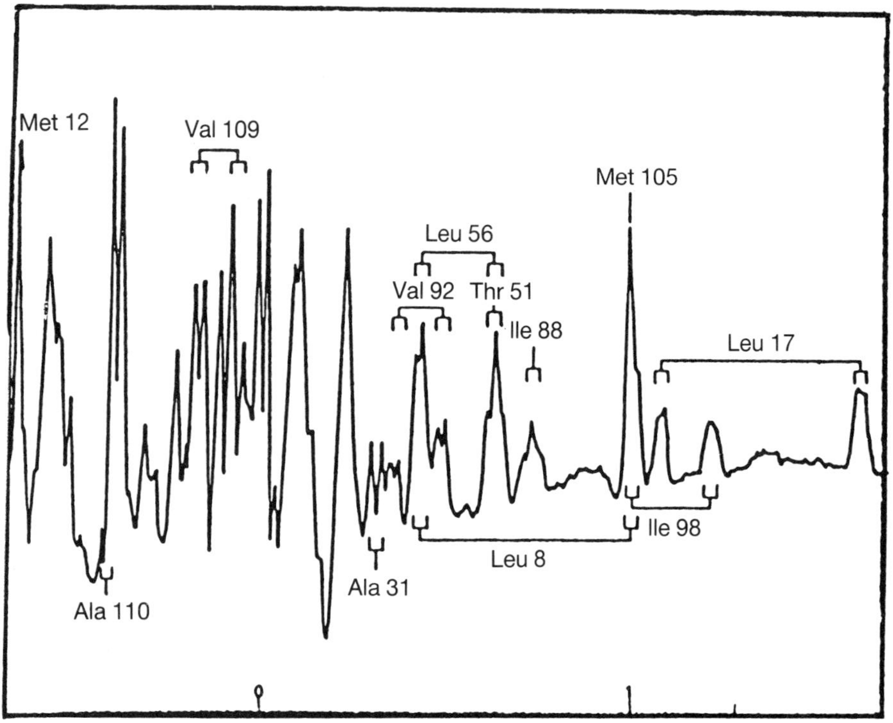

Fig. 2.21 – The assigned residues in a small part of the proton n.m.r. spectrum of lysozyme. The convolution difference spectrum shows the splittings of the resonances clearly. Reproduced from *Chem. Brit.*, **12**, 124 (1976) by permission of the Chemical Society, London.

CYCLIC VOLTAMMETRY

In cyclic voltammetry, the potential of a small, stationary, working electrode (for example Pt) is changed linearly with time, starting from a potential where no electrode reaction occurs, and moving to potentials where reduction or oxidation of a solute takes place. After crossing the potential region in which the electrode reactions occur, the direction of the linear sweep is reversed. The scan rate can be varied over the range 10^2–10^{-5} s. A supporting electrolyte (often n-$Bu_4N^+ClO_4^-$) is present to repress migration of charged products and reactants.

Cyclic voltammetry provides a simple and direct method for measuring the formal potential of a half reaction, when both oxidized and reduced forms are stable during the time required to obtain the voltammogram (current-potential curve). The technique can be illustrated with O_2 (Fig. 2.21). The forward scan begins at an initial potential of −0.75 V and very little current is obtained until about −1.15 V when dioxygen begins to be reduced to superoxide

$$O_2 + e \underset{k_r}{\overset{k_f}{\rightleftharpoons}} O_2^{\dot{-}}$$

The current increases as the rate of reduction increases at more negative potentials, but eventually a maximum is reached (−1.25 V) and the current then decreases steadily. The cathodic peak in the c.v. arises due to competition between two factors: the increase in the net rate of reduction as the potential is made more negative, and the development of a thickening depletion layer across which the reactant must diffuse. The current becomes controlled by the rate of diffusion of reactant through the depletion layer (diffusion-controlled current).

The scan rate is reversed at 1.50 V (switching potential) and the diffusion-controlled current continues until about −1.25 V where net oxidation of $O_2^{\dot{-}}$ back to O_2 occurs. The layer from which O_2 has been depleted is an accumulation layer for $O_2^{\dot{-}}$ and some of this $O_2^{\dot{-}}$ can diffuse back to the electrode and be oxidized. An *anodic peak* in the c.v. is obtained for reasons analogous to those for the cathodic peak.

The c.v. is characterized by several important parameters: the cathodic (E_{pc}) and anodic (E_{pa}) peak potentials, the cathodic (i_{pc}) and anodic (i_{pa}) peak currents, the cathodic half-peak potential ($E_{p/2}$) and the half-wave potential ($E_{½}$). The equation (2.2) is adapted from classical polarography.

$$E_{½} = E^{0'} + (RT/nF)\ln(D_R/D_0)^{½} \tag{2.2}$$

In equation (2.2), $E^{0'}$ is the formal potential related to the ionic strength used, D_0 and D_R are the diffusion coefficients of the oxidized and reduced forms, and n is the number of electrons in the half reaction. Since $D_0 \sim D_R$, $E_{½}$ is usually within a few mV of $E^{0'}$.

The reduction of O_2 is a reversible process. The electron transfer reaction at the electrode surface is so rapid that equilibrium conditions are maintained, even with a substantial net current and a rapidly changing potential. There are several criteria for reversibility; (a) $\Delta E_p = E_{pa} - E_{pc} = 56/n$ mV; (b) $E_{p/2} - E_{pc} = 56.5/n$ mV, values of which must be independent of the scan rate and concentration; (c) the $E_{½}$ is situated exactly (within $2/n$ mV) midway between E_{pa} and E_{pc}, and (d) $i_{pc} = i_{pa}$. Processes which do not obey these criteria are quasi reversible or irreversible. A reaction is quasi reversible if k_f and k_b are of the same order of magnitude over most of the potential range, or totally irreversible if $k_f \gg k_b$ for the cathodic peak and $k_b \gg k_f$ for the anodic peak.

The reduction of O_2 is also diffusion controlled (i.e. no other processes limit the current). The criterion for diffusion control is that $i_{pc}/v^{½}$, where v is the scan rate, must be constant.

POTENTIOMETRIC DETERMINATION OF FORMATION CONSTANTS

The interaction of a metal ion (M) with a ligand (L) can be represented by the series of equilibria

$$M + L \overset{K_1}{\rightleftharpoons} ML; \qquad K_1 = \frac{[ML]}{[M]\,[L]}$$

$$ML + L \overset{K_2}{\rightleftharpoons} ML_2; \qquad K_2 = \frac{[ML_2]}{[ML]\,[L]}$$

$$ML_2 + L \overset{K_3}{\rightleftharpoons} ML_3; \qquad K_3 = \frac{[ML_3]}{[ML_2]\,[L]}$$

$$ML_{n-1} + L \overset{K_n}{\rightleftharpoons} ML_n; \qquad K_n = \frac{[ML_n]}{[ML_{n-1}]\,[L]}$$

where the K values are the *stepwise* formation constants. The *overall formation* constants are defined in terms of the equilibria:

$$M + L \overset{\beta_1}{\rightleftharpoons} ML; \qquad \beta_1 = \frac{[ML]}{[M]\,[L]}$$

$$M + 2L \overset{\beta_2}{\rightleftharpoons} ML_2; \qquad \beta_2 = \frac{[ML_2]}{[M]\,[L]^2}$$

$$M + 3L \overset{\beta_3}{\rightleftharpoons} ML_3; \qquad \beta_3 = \frac{[ML_3]}{[M]\,[L]^3}$$

so that $\beta_1 = K_1, \beta_2 = K_1K_2, \beta_3 = K_1K_2K_3$, etc.

Most ligands L, are the anions of weak acids HL^+, so that formation of a metal complex involves competition between H^+ and M. The most commonly employed method for the determination of formation constants involves potentiometric titration of mixtures of the metal ion and the ligand (for example at 1:1 and 1:2 ratios of metal to ligand) with sodium hydroxide solution. A variety of powerful computer programs is available to deal with the potentiometric titration data such as MINIQUAD, SUPERQUAD, SCOGS, LETAGROP and HALTAFALL. In these programs initial guesses are made as to the values of the formation constants and the constants are refined by an iterative procedure until both the formation and the mass balance equations are satisfied to within a specified tolerance (Table 2.4).

The species present are normally plotted as a function of the pH of the system, and Fig. 2.23 shows plots of the various species for the interaction of Zn(II) with aspartic acid $HO_2CCH_2CH(NH_2)CO_2H$, at 1:1 and 1:2 metal-to-ligand ratios.

It is also possible to study multimetal-multiligand equilibria. For example, the distribution of copper(II) and zinc(II) complexes of the 16 amino acids, $GluNH_2$, Ala, Val, Gly, Pro, Leu, Tyr, Ser, His, iLeu, Orn, Try, Glu, CySSCy, Cys and Met found in blood serum at pH 7.4, 37°C and $I = 0.15$ *M*. The $[Cu^{2+}]$ was 0.018 mM and $[Zn^{2+}]$ = 0.046 mM using the established plasma concentrations of the amino acids (e.g. Ala = 0.38 mM, His = 0.074 mM, Glu = 0.048 mM,

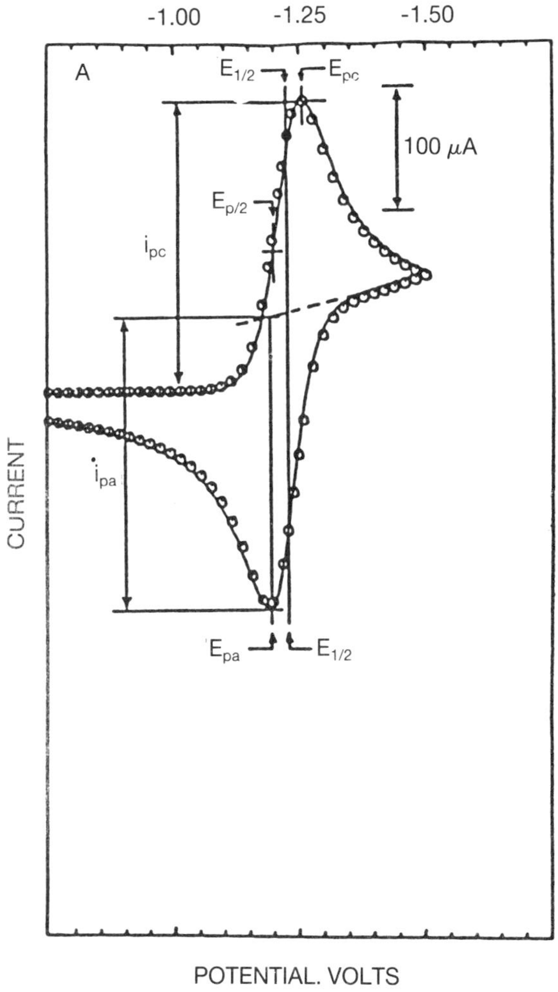

Fig. 2.22 – Cyclic voltammogram of O_2 at a mercury electrode in acetonitrile with a Ag/0.01 M-$AgNO_3$ reference electrode. Supporting electrolyte $(C_2H_5)_4N^+ ClO_4^-$ with O_2 *ca* 1.0 mM. Scan rate 100 mV s^{-1}. $E_{1/2} = -1.22$ V. Reproduced from J. Chem. Ed., **60**, 290 (1983), by permission of the American Chemical Society.

etc). The highest concentrations of the complexes was found to be [Cu(HisH)(CySSCy)] 45.4%, [Cu(His)(CySSCy)] ; 39.6%, $[Cu(His)_2]$ 10.7% and the free metal concentrations $[Cu^{2+}] = 10^{-11}$ M and $[Zn^{2+}] = 10^{-6}$ M.

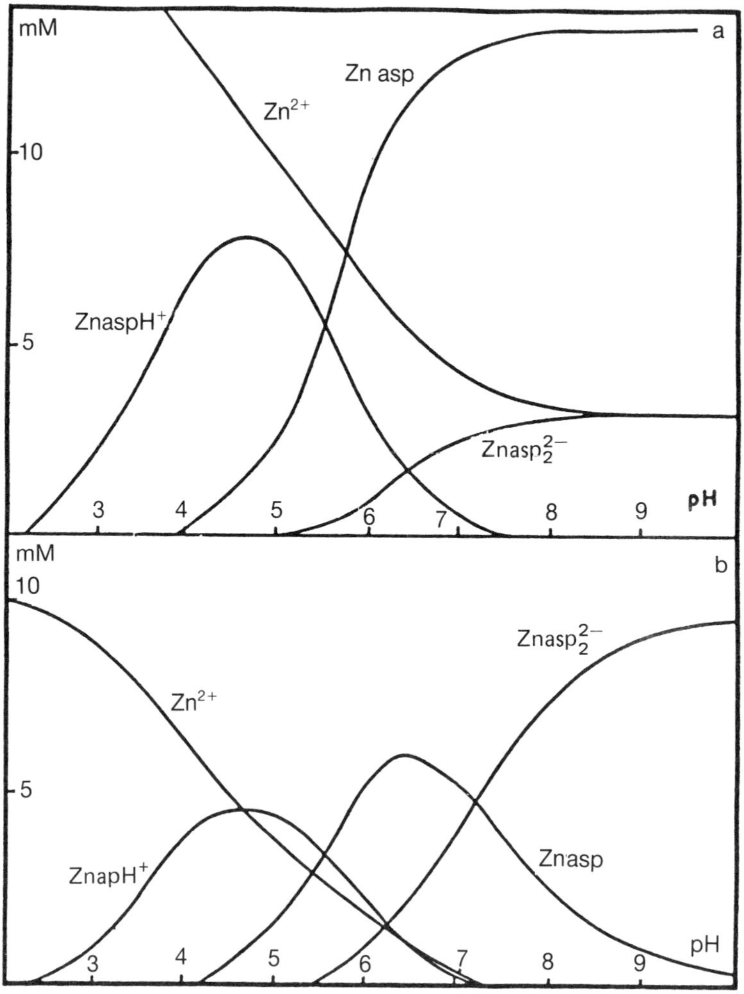

Fig. 2.23 – Species present due to the interaction of zinc(II) and aspartic acid.
(*a*) concentration of aspartic acid $= 2 \times 10^{-2}$ mol dm^{-3}
concentration of zinc(II) $= 2 \times 10^{-2}$ mol dm^{-3}
(*b*) concentration of aspartic acid $= 2 \times 10^{-2}$ mol dm^{-3}
concentration of zinc(II) $= 1 \times 10^{-2}$ mol dm^{-3} asp = (aspartate).
Reproduced from A. M. Fiabane and D. R. Williams "The Principles of Bio-Inorganic Chemistry", by permission of the Chemical Society, London.

Table 2.4 – Summary of some published non-linear least-squares programs for the calculation of formation constants.

Program	Data treated	Sum of squared residuals minimized	Iterative method used
LETAGROP	Potentiometric	Several ($\bar{n}$, analytical hydrogen ion concentration, e.m.f.)	Pitmapping (Newton-Rapheson)
GAUSS	Potentiometric	Analytical hydrogen ion concentration	Gauss-Newton
SCOGS	Potentiometric	Volume of titrant	Gauss-Newton
LEAST	Potentiometric	Analytical concentrations	Gauss-Newton or Newton-Rapheson
LEASK	Potentiometric	Analytical concentrations	Search
STEW	Potentiometric	Analytical hydrogen ion concentration	Fletcher-Powell
MINIQUAD	Potentiometric	Analytical concentrations	Gauss-Newton
SQUAD	Spectrophoto-metric	Absorbance	Gauss-Newton
DALSFEK	Spectrophoto-metric, Potentiometric	Absorbance, e.m.f.	Marquardt

REFERENCES AND BIBLIOGRAPHY

N.m.r.

[1] D. G. Gadian, *Nuclear Magnetic Resonance and its Applications to Living Systems,* The Clarendon Press, New York, 1982.
[2] I. D. Campbell, C. M. Dobson and R. J. P. Williams, *Proc. R. Soc. A,* **23**, 345 (1975).
[3] R. E. Richards, *Endeavour,* **123**, 118 (1975).

O.r.d. and c.d.

[4] R. D. Gillard, *Chem. Brit.*, May, 1867.
[5] J. E. Crooks, *The Spectrum in Chemistry,* Academic Press, London, 1978, p. 155.
[6] P. Crabbé, *ORD and CD in Chemistry and Biochemistry,* Academic Press, New York, 1972.

EXAFS

[7] S. P. Cramer and K. O. Hodgson, 'X-Ray Absorption Spectroscopy, A new Structural method and its applications in bioinorganic chemistry', *Prog. Inorg. Chem.*, **25**, 1-39 (1979).

[8] R. G. Shulman, P. Eisenberger and B. M. Kincaid, 'X-Ray absorption spectroscopy of biological molecules', *Ann. Rev. Biophys. Bioeng.*, **7**, 559-578 (1978).

[9] P. Eisenberger and B. M. Kincaid, *EXAFS, New Horizons in Structure Determination, Science,* **200**, 1441-1447 (1978).

Cyclic Voltammetry

[10] D. H. Evans, K. M. O'Connell, R. A. Petersen and M. J. Kelly, *J. Chem. Ed.,* **60**, 290 (1983). This issue contains a series of articles on various electrochemical techniques, e.g. chronocoulometry and spectroelectrochemistry.

Potentiometric Determination of Formation Constants

[11] F. R. Hartley, C. Burgess and R. Alcock, *Solution Equilibria,* Ellis Horwood, Chichester, 1980.

E.s.r.

[12] N. M. Atherton, *Electron Spin Resonance,* Ellis Horwood, Chichester, 1973.

[13] R. S. Alger, *Electron Paramagnetic Resonance,* Wiley Interscience, New York, 1968.

X-ray Crystallography

[14] F. A. Quicho and W. N. Lipscombe, *Adv. Protein Chem.*, **25**, 1 (1971).

[15] J. Drenth, W. G. T. Hol, J. N. Jansonius and R. Koekock, *Cold Spring Harb. Symp. quant. Biol.*, **36**, 107 (1971).

CHAPTER 3

The Alkali Metal and Alkaline Earth Cations

INTRODUCTION

The bulk metals (Na^+, K^+, Mg^{2+} and Ca^{2+}) constitute about 1% of human body weight, whereas the trace metals (Mn, Fe, Cu, Co, Zn, Mo, V, Cr) represent less than 0.01%. A 70 kg man has approximately 170 g of potassium present in his body (*ca* 9 g in the blood and 3 g in the tissue fluid), but only about 5 g of iron.

Paradoxically it is the trace metals, of the first transition series, which have been most studied during the recent upsurge of interest in the involvement of metal ions in biological processes. The bulk metals have few spectroscopic properties which may be studied, and have therefore remained 'out of sight, out of mind'. The discovery of naturally occurring macrocyclic antibiotics which display ligand and carrier properties towards alkali cations has provided a major impetus to studies of the coordination chemistry of these cations.

During the last decade, three different ligand systems for these cations have been discovered and investigated: (a) the natural acyclic or macrocyclic compounds (cyclodepsipeptides, macrolides); (b) the synthetic macrocyclic polyethers of the 'crown' type, and (c) the synthetic macrocyclic ligands of the 'cryptand' type. Typical examples of these ligands are shown in (3.1)–(3.3).

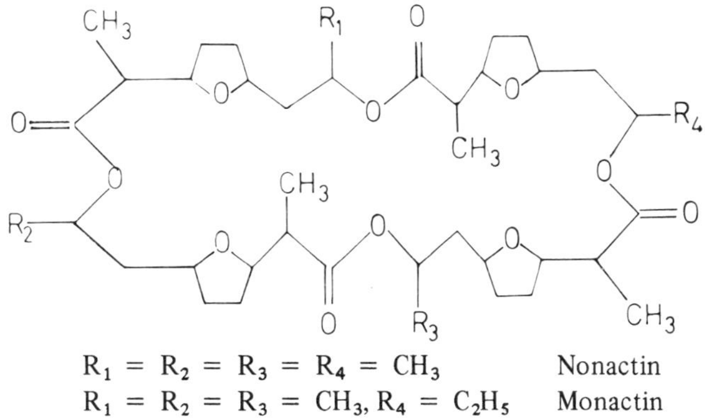

$R_1 = R_2 = R_3 = R_4 = CH_3$	Nonactin	
$R_1 = R_2 = R_3 = CH_3, R_4 = C_2H_5$	Monactin	
$R_1 = R_2 = CH_3, R_3 = R_4 = C_2H_5$	Dinactin	(3.1)
$R_1 = CH_3, R_2 = R_3 = R_4 = C_2H_5$	Trinactin	
$R_1 = R_2 = R_3 = R_4 = C_2H_5$	Tetranactin	

Crown ether

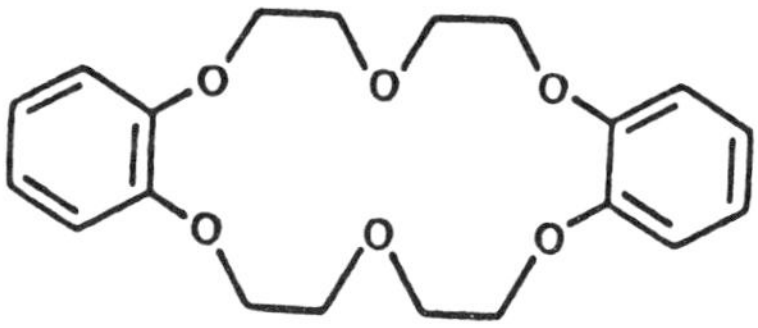

(3.2)

Cryptands

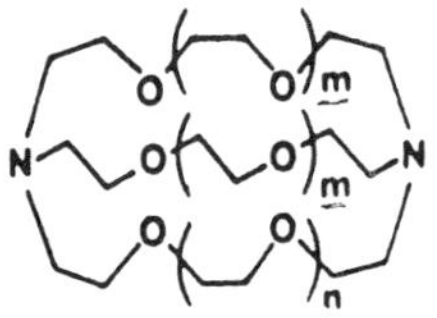

(3.3)

[1.1.1], $m = n = 0$
[2.1.1], $m = 0, n = 1$
[2.2.1], $m = 1, n = 0$
[2.2.2], $m = n = 1$
[3.2.2], $m = 1, n = 2$
[3.3.2], $m = 2, n = 1$
[3.3.3], $m = n = 2$

The first crown compound to be prepared, dibenzo[18]crown-6, (3.2) was observed to be quite insoluble in methanol, but readily dissolved on the addition of sodium salts. This observation led to the discovery of the complexing ability of the crown compounds and to the synthesis of a large range of other macrocyclic polyethers. The nomenclature [18]crown-6 defines the size of the macrocyclic ring and the number of oxygen donors in the ring.

The cryptand ligands contain a molecular cavity (or crypt). The complexes formed are called cryptates. Formation of a cryptate complex can be represented by the equilibrium shown in Fig. 3.1 where K_s is the formation constant of the complex.

The spheroidal intramolecular cavity of macrobicyclic ligands of the cryptand type are particularly well adapted to the formation of stable and selective

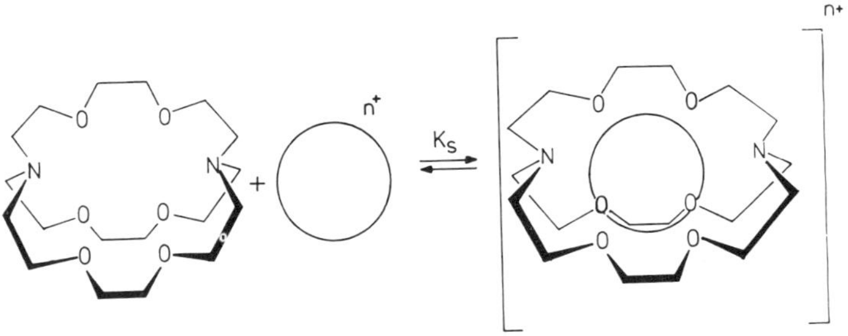

Fig. 3.1 – Formation equilibrium of a cryptate inclusion complex between the macrobicyclic ligand [2.2.2] (taken as an example) and a metal cation; K_s is the formation constant of the complex.

complexes with spherical cations. The complexes may be designated [$M^{n+} \subset L$] where $\subset$ is the mathematical symbol for inclusion and L is the ligand. The cryptands form complexes with suitable alkali and alkaline earth cations which are several orders of magnitude more stable than those of natural macrocycles. For example, [$K^+ \subset$ 2.2.2] is about 10^4 times more thermodynamically stable than [$K^+ \subset$ valinomycin].

THE ALKALI METAL CATIONS

Although sodium and potassium complexes have been known for 50 years or so, and ion pairs have been studied extensively in non-aqueous solution, until recently the aqueous chemistry of these ions had received little attention. Sodium is the principal extracellular cation, and K^+ is the principal intracellular cation. For mammalian blood cells the blood plasma levels are 5 mM (K^+) and 143 mM (Na^+) compared with levels of 105 mM (K^+) and 10 mM (Na^+) for the red blood cells. There is therefore a discriminatory mechanism which controls the selective uptake of K^+ into the cell from plasma.

Some properties of the alkali metal cations are summarized in Table 3.1.

Table 3.1 – Some properties of the alkali metals.

	Li	Na	K	Rb	Cs
Electronic configuration	(He) $2s^1$	(Ne) $2s^1$	(Ar) $4s^1$	(Kr) $5s^1$	(Xe) $6s^1$
Ionization potentials (1)	517	493	416	401	373
(kJ mol^{-1}) (2)	263	4540	3054	2650	2270
Enthalpy of hydration (kJ mol^{-1})	−515	−406	−322	−293	−264
$E°$ (M^+/M) V	−3.02	−2.71	−2.92	−2.99	−3.02
Ionic radius (nm)	0.060	0.095	0.133	0.148	0.174

Representative values of the formation constants of inorganic complexes are shown in Table 3.2. For sulphate, persulphate, thiosulphate, ferricyanide and

Table 3.2 – Stability constants (log β) of inorganic complexes at 25°C and μ = 0.1 M.

Ligand	Li	Na	K	Rb	Cs
Sulphate	0.64	0.70	0.82	–	–
Persulphate	–	0.58	0.91	1.17	1.42
Thiosulphate	–	0.58 ± 0.02	1.00 ± 0.04	–	–
Ferrocyanide	1.78	2.08	2.3	2.65	2.85
	–	–	2.35 ± 0.2	–	–
Ferricyanide	–	–	1.46 ± 0.02		
	–	−0.32 ± 0.13	0.30 ± 0.04	0.52 ± 0.03	–
Carbonate	–	0.55	–	–	–
Bicarbonate	–	0.16	–	–	–
Phosphate	0.72 ± 0.04	0.59 ± 0.04	0.48 ± 0.06	–	–
Trimetaphosphate	–	0.88	no ev.	–	–
Tetrametaphosphate	–	1.42	1.26	–	–
Triphosphate	2.87 ± 0.06	1.64 ± 0.06	1.37 ± 0.06	–	–
Tetraphosphate	2.64	1.79	1.71	–	–

ferrocyanide, the formation constants decrease in the order Cs > Rb > K > Na > Li, whereas phosphates show the opposite preference, although the difference between the stabilities of potassium, rubidium and caesium complexes is quite small. Complex formation is favoured by a high charge on the ligand and by the ease with which chelation can occur. For example, cyclic condensed phosphates give more stable complexes than linear condensed phosphates.

In 1964, the American physiologist B. C. Pressman found that certain antibiotics could induce the selective movement of K^+ into rat liver mitochondria. These antibiotics, now collectively termed ionophoric (ion-bearing) antibiotics could also increase the permeability of black-lipid films (synthetic lipid bilayers) to K^+. The antibiotics are neutral at physiological pH and act as discriminatory cation carriers by forming complexes with the alkali metal cations.

MEMBRANES

The cell is surrounded by a membrane which separates its aqueous interior from the plasma, and selectively allows into the cell ions and nutrients, and allows out unwanted material or material produced for use elsewhere. The membrane functions as a 'living barrier' and it is in this barrier that the process of cation discrimination occurs.

The membrane is about 70 Å thick and is composed of protein and lipids. A recent view of the membrane (the 'Fluid Mosaic Model') considers that it consists of a lipid bilayer in which globular proteins float 'as icebergs' (Figs. 3.2 and 3.3). The alkali metal cations must traverse this barrier passing through a medium of low dielectric constant. This passage is not favoured in view of the large electrostatic energy required to transfer an ion from a region of high dielectric constant (the plasma) into one of low dielectric constant (the lipid bilayer).

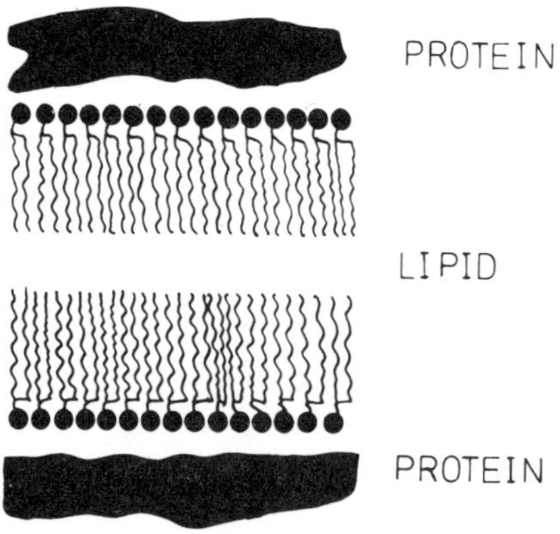

Fig. 3.2 – The trilamellar structure of the membrane. Reproduced from *Chem. Soc. Rev.*, **6**, 325 (1977) by permission of the Chemical Society, London.

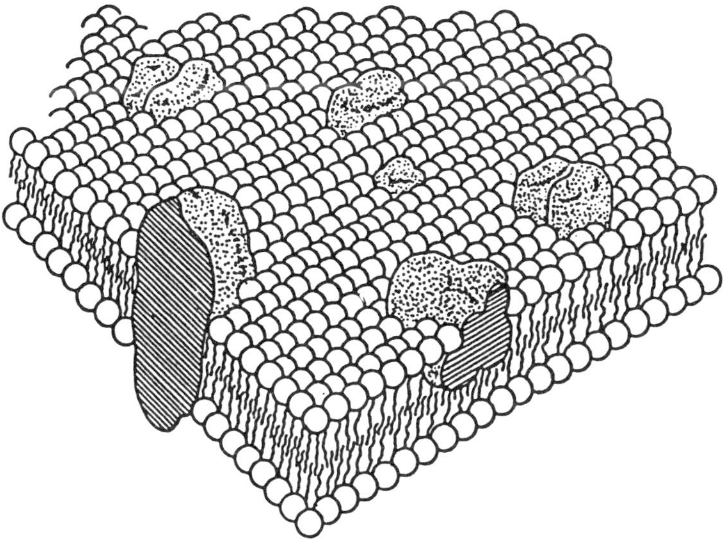

Fig. 3.3 – The Fluid Mosaic Model for the membrane [Reproduced by permission from *Science*, **175**, 720 (1972)].

A carrier molecule is required to encapsulate the cation, thus presenting an organic, lipid-soluble surface to the membrane. A suitable ligand can provide an 'organic overcoat' for the cation. The alkali cation may cross the membrane by being passed from donor site to donor site in a pore.

There can be carrier-assisted passive transfer of material independent of an energy source, the so-called facilitated diffusion. Or the cell can accumulate cations by working against a concentration gradient. The second process requires energy and is known as active transport. The hydrolysis of adenosine triphosphate (ATP) to adenosine diphosphate (ADP) and inorganic phosphate is believed to provide the energy source for this process, generally known as the 'sodium pump'. Model studies with liquid membrane systems have been carried out, using the U-tube technique. A system is constructed using a U-tube consisting of two aqueous layers separated by a semipermeable medium, or membrane. Chloroform ($CHCl_3$) is chosen because its dielectric constant is similar to that of the membrane. A coloured alkali-metal salt (e.g. potassium 4-nitrophenolate) is introduced on one side of the barrier and this is set aside as a control experiment (Fig. 3.4a). A second tube is prepared with a carrier ligand in the $CHCl_3$ layer. The movement of colour on transport by the ligand carrier can then be noted. If the colour transfers rapidly into the organic layer, but no further, the ligand is acting as a ion receptor (Fig. 3.4b). If the colour is readily transferred through the $CHCl_3$ layer and into the second aqueous layer the added molecule is acting as an ion carrier. The difference in behaviour is related to the formation constant of the complex. A high formation constant leads to ion reception and a medium value to ion carriage.

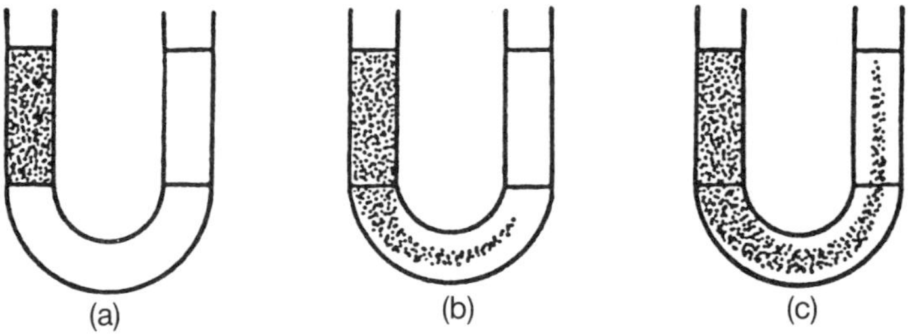

Fig. 3.4 – The U-tube experiment: the shaded portions represent the colour imparted to the solvent phases by the potassium *o*-nitrophenolate; the right hand columns contain water. Reproduced from *Chem. Soc. Rev.*, **6**, 325 (1977) by permission of the Chemical Society, London.

The best carrier for ion transport is a ligand which gives a moderately stable rather than a very stable complex. A very stable complex cannot release the ion efficiently from the complex. The cyclic dodecadepsipeptide valinomycin has a high ion selective complexation of K^+ versus Na^+. Enniatin B is a cyclohexadepsipeptide which has a reduced K^+/Na^+ selectivity (for the 1:1 complex) relative to valinomycin, which is ascribed to the greater flexibility of the latter ligand. The structure of the 1:1 enniatin B-KI complex is shown in Fig. 3.5.

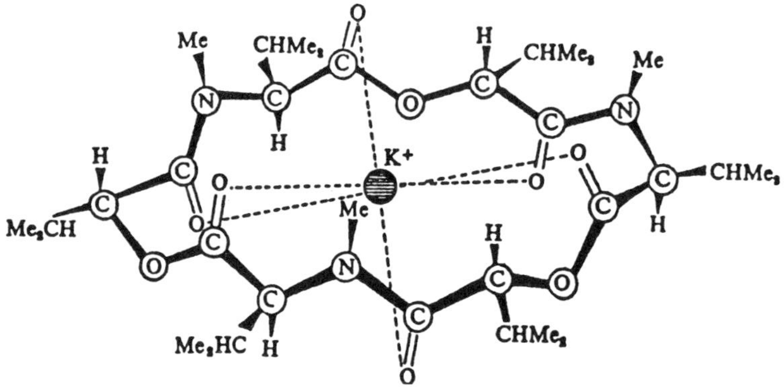

Fig. 3.5 – The enniatin B-potassium complex.

CALCIUM PROTEINS

The unique biological role of Ca^{2+} can be appreciated by Ringer's observation (1882) that isolated turtle hearts would continue beating for many hours, if and only if, 10^{-3} M Ca^{2+} was present in the isotonic bathing medium.

Studies of the Ca^{2+} complexes of small ligands have established that oxygen is the preferred donor atom even though a nitrogen donor may be available in the ligand. The coordination number is usually eight, but seven-and six-coordinate complexes are known. For eight-coordination the geometry is best described as a square antiprism. Seven coordinate complexes are pentagonal bipyramidal, and six-coordinate Ca^{2+} is a near regular octahedron. Calcium–oxygen bond distances can be correlated with the coordination number (eight-coordinate 2.452 Å, seven-coordinate 2.39 Å and six-coordinate 2.33 Å).

A number of calcium–protein complexes are known such as parvalbumin, troponin, concanavalin A, staphylococcal nuclease and thermolysin.

Parvalbumin

Muscle calcium binding parvalbumin (MCBP) is found in the white muscle of chordates. The protein has a molecular weight of about 11,500 and binds two Ca^{2+} ions ($pK_d = 6.7$, where K_d is the dissociation constant). The crystal structure has been refined to 1.9 Å resolution and the structure has been shown to consist of six α-helical regions denoted A–F. One Ca^{2+} is bound in the loop between helices C and D, and the second is bound in the EF region, referred to as the 'EF hand'. The CD calcium ion is six-coordinate with oxygen donors from Asp-51, Asp-53, Ser-55, Phe-57, Glu-59 and Glu-62. The EF calcium ion is bound by five amino acids (Asp-90, Asp-92, Asp-94, Lys-96 and Glu-101) and by water. This calcium is formally eight-coordinate as the carboxylate groups of Asp-92 and Glu-101 donate via both oxygens (the Phe-57 and Lys-96 peptide oxygens coordinate calcium).

Troponin

The troponin complex of muscle is a 'trimer' of three separate proteins one of which, troponin-C (TN-C), binds four Ca^{2+} ions with pK_d values in the range 5.5 to 7.5. The main ligating oxygen atoms are derived from the carboxylate groups of Glu and Asp amino-acid residues, however, two hydroxyl oxygens of serine (Ser-35 and Ser-67) and the amide oxygen of asparagine-104 are also involved.

Staphylococcal nuclease

This enzyme hydrolyses both DNA and RNA to 3′-mononucleotides. The enzyme binds a single Ca^{2+} with an apparent pK_d of 3. The calcium is essential for substrate binding and enzymic catalysis. The crystal structure to nearly 2 Å resolution has been determined by Cotton and coworkers. Calcium coordination is approximately octahedral, from the carboxyl groups of Glu-43, Asp-19, Asp-40 and the carbonyl oxygen (peptide bond) of Thr-41. A sixth ligand, possibly an water molecule, occurs and both oxygens of Asp-21 may be involved in coordination.

Thermolysin

Thermolysin is a heat stable, neutral protease from *Bacillus thermoproteolyticus* with a molecular weight of = 37,500 and containing four Ca^{2+} and one Zn^{2+}. If three or four Ca^{2+} ions are removed, the catalytic activity is not greatly altered, but the protein becomes quite heat labile. Calcium is apparently involved in maintaining the quaternary structure of the enzyme.

Concanavalin A

Concanavalin A (subunit of molecular weight 25,200) exists as a dimer between pH 3.5 and 5.6. Con-A binds carbohydrates and its biological effects are probably associated with its sugar binding properties. The protein binds both Ca^{2+} and Mn^{2+} at sites over 20 Å distant from the carbohydrate binding site. The metal ions appear to be essential for sugar binding.

CALCIUM FUNCTIONS

The concentrations of Ca(II) inside and outside of the cell are approximately 10^{-6} mol dm^{-3} and 10^{-3} mol dm^{-3} respectively. A range of biochemical and/or physiological processes are triggered off by entry of Ca(II) into the cell, or by release of Ca(II) from internal organelles.

For various types of muscle, the sarcoplasmic reticulum is well established as the Ca(II) regulator. Calcium(II) is taken up by the Ca(II), Mg(II)-ATPase with a number of similarities to Na(I), K(I)-ATPase, which is present in the sarcoplasmic reticulum (SR) membrane. The Ca(II) is stored in the SR on calcium-binding proteins, particularly calsequestrin, which has a high affinity for Ca(II) and binds some 43 moles of Ca(II) per mole of protein (M = 33,000). Muscle contraction is associated with the release of Ca(II) from the SR, and its binding to sites on the muscle fibres. The concentration of Ca(II) in the sarcoplasm rises 100-fold in milliseconds.

Some of the best understood Ca(II)-binding regulatory proteins are found in muscle. Initiation of contraction results from the arrival of a nerve impulse at a motor ending in a fibre, which causes Ca(II) to be released from its stores. This Ca(II) then interacts in a specific manner with a regulatory protein, although this varies with the type of muscle. In vertebrate skeletal muscle, Ca(II) interacts with troponin, but in fish and amphibian muscle, the sarcoplasm contains large amounts of parvalbumins, which are water soluble, low molecular weight proteins which bind Ca(II), and are probably equivalent in overall function to troponin. The binding of Ca(II) to troponin-C has been much studied, and two high-affinity and two low-affinity sites on the protein have been characterized. High resolution proton n.m.r. measurements show that troponin-C exists in three conformational states, corresponding to the binding of 0, 2 and 3 moles of Ca(II).

Calcium and Secretion

In many secretory cells, materials such as hormones, neurotransmitters and other control and defence mechanisms, are stored in vesicles or granules. After an appropriate stimulus, these granules migrate to the periphery of the cell and are extruded in the process of exocytosis. Calcium is implicated at some stage of the stimulus-secretion coupling. Examples of secretion by excocytosis include histamine (from mast cells) and insulin (from pancreatic cells). All these processes involve the expenditure of energy and the presence of calcium.

Calcium and blood clotting

Coagulation of blood occurs to prevent excessive bleeding when tissues are damaged. The overall mechanism is complex and involves a cascade process, many steps of which are dependent on calcium. This cascade process (a set of coupled reactions) involves a number of proteins which are normally present in blood as inactive or precursor forms.

Calcification

The deposition of calcium salts is an essential feature of the development of extracellular structures such as shell, bones and teeth. Deposition in the incorrect location can lead to stone formation, osteoarthritis, cataracts and arterial disorders. Complex systems are available for the control of the mobilization and deposition of calcium, and include parathyroid hormone, vitamin D, calcitorin, calsequestrin and osteocalcium. Calcium is often stored in the tissues as granules, which can be mobilized for shell formation and other processes. The calcium salts are present as small crystals so that deposition and reabsorption rates will be rapid and equilibria will be maintained.

Calmodulin

It is becoming clear that the calcium-binding regulatory proteins found in many Ca(II) dependent processes are identical. This protein has been called 'Clamodulin', in view of its role as the intracellular calcium receptor.

Magnesium

Magnesium is the fourth most abundant cation in the body, about one-half being present in the skeleton. It is essential for life and plays a major role in the regulation of a number of vital biological processes, including enzymic reactions. Plasma magnesium is regulated within fine limits, although the control mechanism is not fully established. It has been suggested that magnesium ions must have been very important in very early life processes. Calcium ions, however, acquired their functions of specific importance much later. The calcium functions involve processes that only characterize the higher forms of life, such as nerve transmission, muscle contraction and blood coagulation. In blood plasma and other biological fluids, calcium ions predominate, and the concentration of Mg^{2+} is relatively low. However, within cells, the concentration of Mg^{2+} is high, while the concentration of Ca^{2+} is low. There is a pronounced gradient in the Ca^{2+} concentration across the cell membrane, and this is important for the biological action of Ca^{2+}. A further difference between Mg^{2+} and Ca^{2+} occurs at the molecular level. In enzymic processes, such as protein biosynthesis and anaerobic energy production, Mg^{2+} reacts with the substrate rather than the enzyme. The larger Ca^{2+}, having a higher coordination number (six, seven or eight), usually binds directly to the enzyme molecule, and can induce substantial conformational changes, for example, in the Ca^{2+} activation of staphylococcal nuclease.

The calcium ion (r = 0.99 Å) has a larger ionic radius than the magnesium (r = 0.65 Å), so that Ca^{2+} has a much lower charge density. Substitution reactions on Ca^{2+} are therefore much faster than those of Mg^{2+}. These properties cause Ca^{2+} to move quite rapidly in living tissues. In higher organisms Ca^{2+} has been developed as a 'trigger' for transferring signals between different cells (e.g. muscle contraction).

Magnesium(II) can substitute for Mn(II) in a number of enzymic reactions. Hence chickens grown on a manganese-free diet synthesize an oxaloacetate decarboxylase containing Mg(II) rather than Mn(II). Magnesium(II) is involved in the hydrolysis of many biologically important phosphate derivatives such as ATP → ADP + P_i. Complex formation involving two macromolecules, for example, interaction of two nucleic acids, or the interaction of nucleic acids with protein molecules would lead to disorder within the cell (ΔS large and positive), due to the release of water molecules, unless the nucleic acids were present as complexes with Mg^{2+}.

BIBLIOGRAPHY

Polyethers and their Complexes

[1] C. J. Pedersen and H. K. Frensdorf, 'Macrocyclic Polyethers and their Complexes', *Angew. Chem., Internat. Edit.*, **11**, 16 (1972).

Crytates

[2] J.-M. Lehn, 'Cryptates: The Chemistry of Macropolycyclic Inclusion Complexes', *Acc. Chem. Res.*, **11**, 49 (1978).
[3] J.-M. Lehn, *Struct. Bonding (Berlin)*, **16**, 1 (1973).

General

[4] R. J. P. Williams, 'The Biochemistry of Sodium, Potassium, Magnesium and Calcium', *Quart. Rev.*, **24**, 331 (1970).
[5] D. E. Fenton, 'Across the Living Barrier', *Chem. Soc. Rev.*, **6**, 325 (1977).
[6] Yu. A. Ovchinnikov, V. T. Ivanov and A. M. Shkrob, *Membrane Active Complexones*, Elsevier. Amsterdam, 1974.

[7] R. Harrison and G. C. Lunt, *Biological Membranes,* Blackie, Glasgow and London, 1975.
[8] M. Dobler, *Ionophores and their Structures,* John Wiley, Chichester, 1981.
[9] D. Midgley, 'Alkali-metal Complexes in Aqueous Solution', *Chem. Soc. Rev.,* **4**, 549 (1975).

Calcium

[10] C. P. Bianchi, *Cell Calcium,* Butterworths, London, 1968.
[11] C. J. Duncan, ed., 'Calcium in Biological Systems', *Soc. Exp. Biol. Symp.,* XXX, 1976.
[12] R. H. Wasserman *et al.,* eds., *Calcium Binding Proteins and Calcium Function,* North Holland, 1977.
[13] Calmodulin, A. R. Means and J. R. Dedman, *Nature,* **285**, 72 (1980).

CHAPTER 4

Non-Redox Metalloenzymes

INTRODUCTION

As discussed in Chapter 1, metalloenzymes can be defined as enzymes with strongly bound metal ions in which the metal ion is an essential participant in the catalytic process. Redox enzymes are considered in Chapters 6 and 7 and the present chapter will concentrate on metalloenzymes in which the metal ion acts primarily as a Lewis acid.

ZINC(II)

Zinc(II) plays an important role as a Lewis acid in many metalloenzymes. The widespread use of zinc(II) as a Lewis acid probably arises for the following general reasons:

(a) Zinc(II) is a d^{10} ion, and therefore has no ligand field effects associated with it to determine a particular coordination number or geometry.
(b) Ligand exchange processes on zinc(II) are rapid, so that substrates and products can be readily introduced and removed.
(c) Zinc(II) is a reasonably effective Lewis acid and is active in many model systems.
(d) Zinc(II) does not hydrolyse to form hydroxo complexes at low pH. The pK_a of the aquo zinc(II) ion is *ca* 8.8.
(e) Zinc(II) has no redox chemistry associated with it under biological conditions

MANGANESE(II)

Manganese(II) can also act as an effective Lewis acid in such enzymes as oxaloacetate decarboxylase and pyruvate carboxylase. In some cases Mg(II) can substitute for manganese(II).

BIOLOGICAL LEWIS ACIDS

Biological systems have chosen to use Mg(II), Ca(II), Mn(II), Ni(II) and, first and foremost, Zn(II) to fulfil the role of Lewis acids in metalloenzymes. It is interesting that although Cu(II) plays a major role in many electron-transfer proteins it does not appear to function in any non-redox metalloprotein. In model systems, Cu(II) is the most active Lewis acid. The simplest rationale for the selection of metal ions for many metalloenzymes, appears to be that those metal ions which can readily take part in redox reactions (under normal biological conditions), are excluded from those proteins in which they would be required to have a role related to their Lewis acidity. The redox chemistry of Mn(II), Zn(II), Ni(II) and Ca(II) does not readily occur under normal biological conditions.

APOENZYMES

The *in vitro* preparation of apoenzymes by the removal of the native metal ion from metalloenzymes by various chelating agents has been an important part of of the study these catalysts. Clearly, the preparation of a inactive metal-free apoenzyme, which can be completely reactivated by the addition of an appropriate metal ion, confirms the metal ion requirements of the enzyme system.

For the three zinc(II) metalloenzymes (carboxypeptidase A, carbonic anhydrase and alkaline phosphatase), stable apoenzymes have been prepared and their properties studied in detail. It is often possible to introduce other metal ions into the apoenzyme, and cobalt(II) carbonic anhydrase has not a too dissimilar catalytic activity to the native zinc(II) enzyme. Such metal substitution reactions can provide very useful information. Zinc(II) is a d^{10} ion and is spectroscopically 'silent' i.e. lacks for example e.s.r. and d–d spectra. The introduction of Co(II) into the apoenzyme allows details of the active site to be probed by spectroscopic techniques (Chapter 2).

METAL-ION CATALYSED REACTIONS

Aquo ions catalyse a variety of organic reactions in solution (Table 4.1). In fact for every proton catalysed reaction (specific acid catalysis), a metal-ion catalysed analogue can be developed. As specific acid catalysis is generally inefficient at pH 7, enzymes make use of general acid catalysis and metal ion catalysis. Enzymic reactions are usually at least 10^9-fold faster then the uncatalysed reaction and many metal ion catalysed reactions are known with rate accelerations of 10^6. Thus the hydrolysis of ethyl glycinate in the copper(II) complex (**4.1**) has $k_{OH} = 1.4 \times 10^5\ M^{-1}\ s^{-1}$ at 25 °C compared with $k_{OH} = 0.63\ M^{-1}\ s^{-1}$ for the hydrolysis of $NH_2CH_2CO_2Et$ at 25 °C, a rate acceleration of 2×10^5.

Table 4.1 – Reactions catalysed by metal ions acting as Lewis Acids.

Reaction	Reaction
Ester hydrolysis	Glycoside hydrolysis
Amide hydrolysis	Acetal hydrolysis
Peptide hydrolysis	Hydrogen exchange
Phosphate ester hydrolysis	Sulphate ester hydrolysis
Schiff base hydrolysis	Fluorophosphate hydrolysis
Carbonyl hydration	Nitrile hydrolysis
Peptide bond formation	Schiff base formation
Transamination	Transphosphorylation
Carboxylation	Thiol ester hydrolysis
Decarboxylation	Hydride reductions

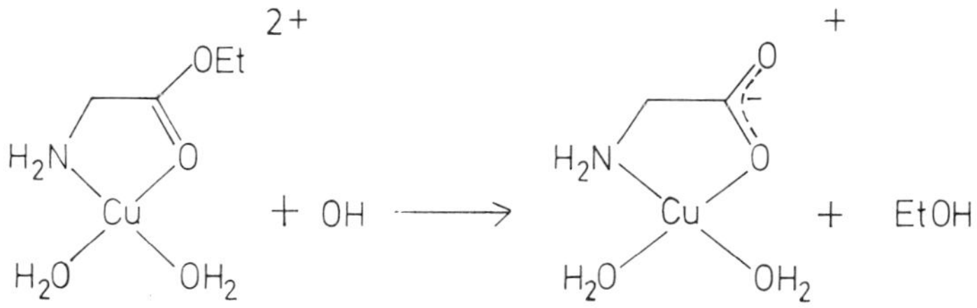

(4.1)

COMPARISONS OF A METAL CATALYSED REACTION AND AN ENZYMIC REACTION

It is of interest to compare a simple metal ion catalysed reaction, with the corresponding metalloenzyme catalysed reaction in which the metal ion acts as a Lewis acid.

The decarboxylation of oxaloacetic acid to give pyruvic acid and CO_2,

$$HO_2CCOCH_2CO_2H \longrightarrow HO_2CCOCH_3 + CO_2$$

is catalysed by many cations, and by the manganese(II) metalloenzyme oxaloacetate decarboxylase.

Catalysis by metal cations involves the sequence of reactions shown in Scheme 4.1. The rate constant for the decarboxylation of $(ZnA)_{ketonic}$ is $7.42 \times 10^{-3}\ s^{-1}$ at 25 °C, while for $(CuA)_{ketonic}$, $k = 0.17\ s^{-1}$. The copper(II) complex decarboxylates some 23 times faster than the zinc(II) complex. The rate constant for the decarboxylation of the dianion A^{2-} is $1.7 \times 10^5\ s^{-1}$ so that the Zn(II) complex decomposes 436 times faster than A^{2-}. Although Cu(II) is a more effective Lewis acid than Zn(II) in assisting the transfer of electrons from the carbon-carbon bond undergoing cleavage, the effects of the two metal ions are not greatly different.

The metal-ion catalysed reaction shows a normal ^{13}C-^{12}C carbon isotope effect of some 6%, indicating that cleavage of the C-C bond is rate-determining. The enzymic reaction does not display a carbon isotope effect, but the reaction is slower in D_2O than in H_2O, whereas the rate of the Mn(II)-catalysed reaction is unaffected by the change of solvent.

It is clear that in the enzymic reaction, the rate determining step is not cleavage of the carbon-carbon bond, which is considered to occur some 10^8 times faster in the enzymic reaction than for the metal-ion catalysed reaction. The rate determining step occurs in a subsequent process involving proton transfer, such as ketonization of the enol intermediate and release of pyruvic acid from the enzyme.

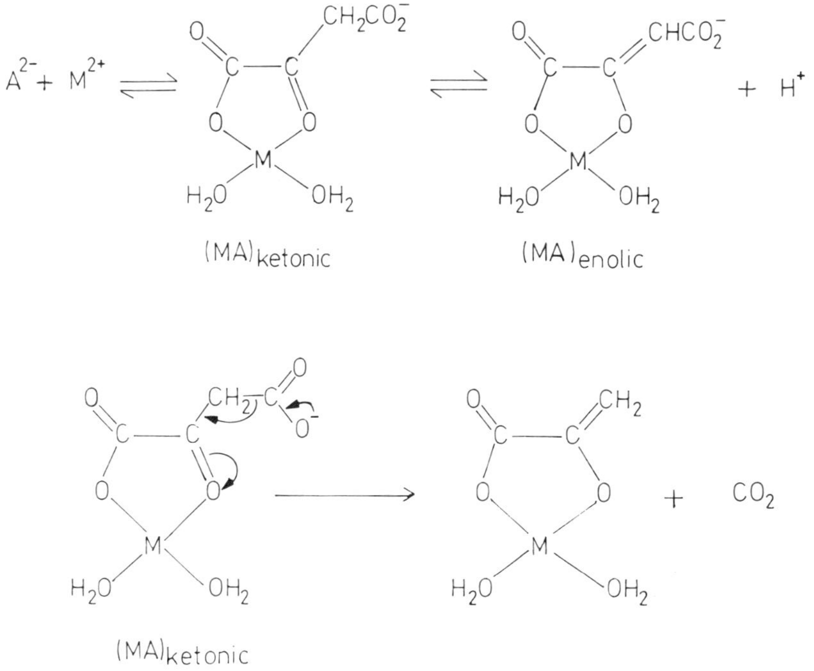

Scheme 4.1 – The metal-ion promoted decarboxylation of oxaloacetic acid (A^{2-} = the dianion of OAA).

LEWIS ACID CATALYSIS

The above example illustrates a number of important general points:

(a) There is not a marked difference in the Lewis acidity of metal ions of the second half of the first transition series. Lewis acidity will depend to a large degree on the ratio of charge to size of the metal cation. (Mn(II) = 82 (HS), Co(II) = 65 (LS), 73 (HS), Ni(II) = 70 (HS), 61 (LS), Zn(II) = 74 are the ionic radii in pm, HS = high spin; LS = low spin).

(b) Enzymic reactions may occur by a similar *general* mechanism to simple metal-ion promoted reactions, but may involve changes in the rate determining step.

(c) The Lewis acidity of a metal ion can probably be 'tailored' to a degree by the ligand groups from the protein. Ligands are Lewis bases, and as a result the stronger the σ-electron donation from the ligand to the metal, the weaker the Lewis acidity of the metal centre. Ligands such as imidazole which are both σ-donors and π-acceptors may tailor the metal ion's Lewis acid character by the appropriate degree of π-acceptance.

In many metal-ion catalysed reactions, the rate acceleration which can be as high as 10^6-fold arises primarily from the $\Delta S^{\ddagger}$ term. The hydrolysis of esters, amides and peptides normally involves the formation of a tetrahedral intermediate which then breaks down to products.

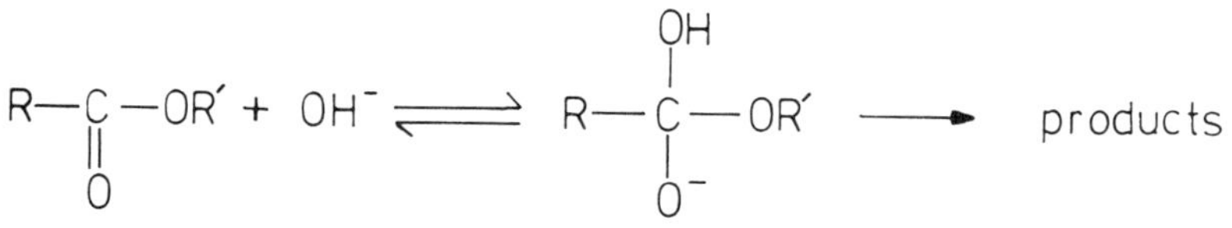

The developing negative charge on the carboxyl oxygen atom requires orientation of solvent in this area of the molecule, which contributes to the overall large negative $\Delta S^{\ddagger}$ of a bimolecular reaction (*ca* -80 JK^{-1}mol^{-1}). In the metal-ion catalysed process,

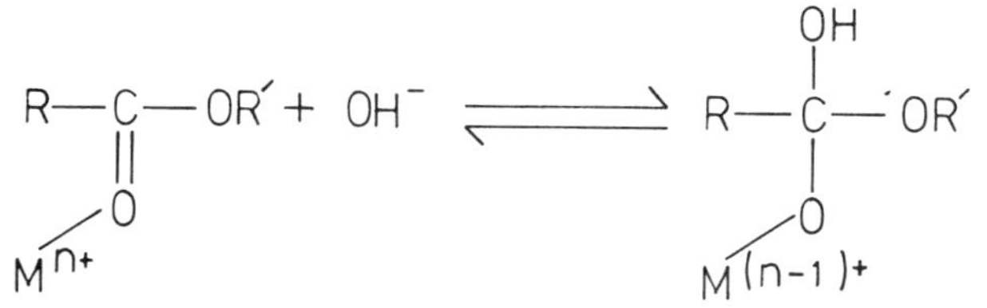

the metal ion effectively 'solvates' the developing negative charge making $\Delta S^{\ddagger}$ *less* negative. The effect of changes in the enthalpy of activation are summarized in Table 4.2. Rate accelerations of 10^6-fold can arise if $\Delta S^{\ddagger}$ is made positive by *ca* 125 JK^{-1}mol^{-1}.

Table 4.2 – The effect of changes of $\Delta S^{\ddagger}$ on rate constants at a constant enthalpy of activation.

$\Delta(\Delta S^{\ddagger})$ JK^{-1}mol^{-1}	k/k_0
4.18	1.64
8.36	2.72
20.9	12.2
41.8	148
83.6	21,900
125.4	3.2×10^6

CARBOXYPEPTIDASE

Two classes of carboxypeptidase have been recognized. Enzymes of one class, for example, yeast carboxypeptidase C are intracellular, showing maximal activity at acidic pH. They are *not* metalloenzymes and their mechanism of action may be similar to that of the serine proteases such as α-chymotrypsin. The second class of carboxypeptidases are metalloenzymes which are released from their inactive precursors (zymogens) in the pancreatic juice of animals. These enzymes exhibit maximal activity at neutral or slightly alkaline pH and act extracellularly, aiding protein digestion in the duodenum.

Carboxypeptidases catalyse the hydrolysis of the C-terminal amino-acid residue from a peptide or protein chain. Carboxypeptidase is an example of an exopeptidase which cleaves only terminal peptide bonds. Endopeptidases catalyse the hydrolysis of peptide bonds which are non-terminal. The absolute requirements for carboxypeptidase A are that the C-terminal residue must have the S-configuration and that its carboxyl group must be free. Substrates in which the amino-acid side chain is aromatic are favoured, but carboxypeptidase A has a fairly broad specificity in this respect. Peptides possessing almost any C-terminal residue except proline will be hydrolysed.

Bovine carboxypeptidase A and B have been sequenced, and their structures established by X-ray techniques. Carboxypeptidase A (CPA) consists of a single peptide chain of 307 amino-acid residues (M = 34,300) and contains one atom of zinc. Carboxypeptidase B (CPB) has 308 residues, 49% of its sequence being identical to that of CPA and it also contains one atom of zinc. The three-dimensional structures of the two enzymes are similar. Carboxypeptidase A is ellipsoidal (50 × 40 × 38 Å). The zinc(II) ion in carboxypeptidase A may be removed either by dialysis at low pH or by dialysis at neutral pH against buffer containing the chelating agent, 1,10-phenanthroline. The loss of activity from samples treated in this manner exactly parallels the loss of zinc. The removal of zinc appears to have little effect on the overall structure of the protein. The optical rotations of the native and metal-free enzymes and their behaviour on sedimentation are identical. Addition of zinc to the apoenzyme restores the catalytic activity.

Certain other transition metal ions may replace zinc(II) and regenerate peptidase activity, for example Fe(II), Mn(II), Co(II) and Ni(II). These metal ions occupy the site formerly occupied by the Zn(II) ion. The cobalt(II) enzyme is reported to be a better catalyst in the hydrolysis of *N*-Cbz-glycylphenylalanine than the native enzyme (Table 4.3).

Table 4.3 – Relative apparent activities of some metallocarboxypeptidases*

Metal	Peptidase activity	Esterase activity
apo (no metal)	0	0
Zn	1.0	1.0
Co	1.6	0.95
Ni	0.08	0.35
Cd	0	1.50
Hg	0	1.16
Pb	0	0.52
Ca	0	0

*From data of Coleman and Vallee. *J. Biol. Chem.*, **236**, 2244 (1961).

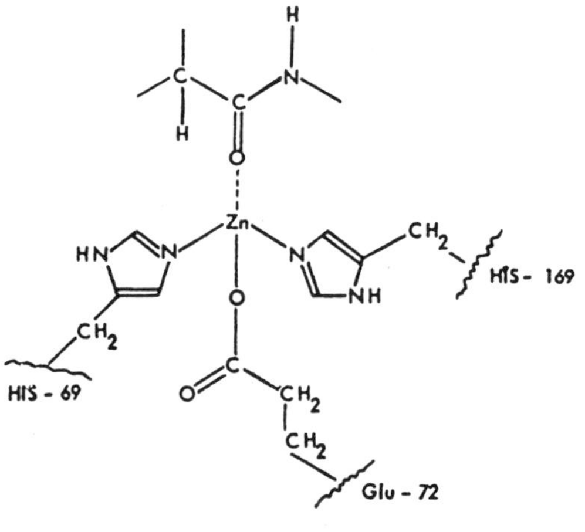

Fig. 4.1 – The coordination of the zinc(II) ion at the active site of carboxypeptidase A.

The active sites of CPA and CPB are very similar, and the coordination of zinc(II) at the active site of carboxypeptidase A is shown in Fig. 4.1.

Each enzyme's zinc(II) is coordinated in a distorted tetrahedral stereochemistry to two histidine residues (His 69 and His 169), one glutamic acid residue, and a water molecule (or HO^-). A substrate molecule displaces the water ligand of the zinc with the peptide oxygen coordinating to zinc. The crystal structure of CPA containing glycyl-L-tyrosine (a poor substrate which binds but is not hydrolysed) shows that the amide carbonyl group is ligated to zinc. The coordinated carbonyl group is polarized and attacked by a water molecule activated by Glu-270. The resulting tetrahedral intermediate is protonated on nitrogen by Tyr-248 and can then decompose to amine and carboxylic acid (Fig. 4.2).

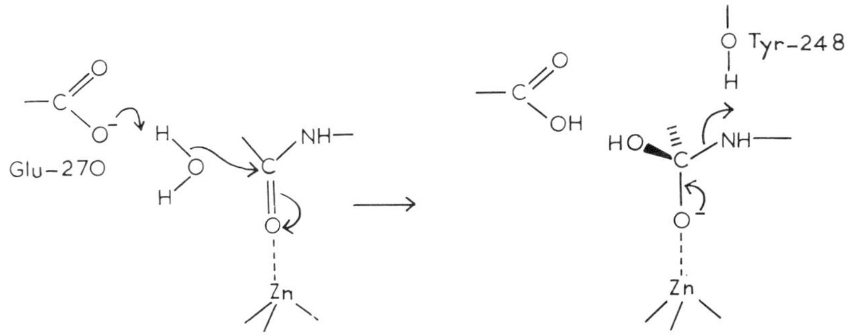

Fig. 4.2 – Zn-carbonyl mechanism for CPA (peptide substrate), involving Glu-270 as a general base.

An alternative mechanism involves Glu-270 acting as a nucleophile rather than as a general base. In this case a mixed anhydride intermediate is formed which is subsequently hydrolysed (Fig. 4.3).

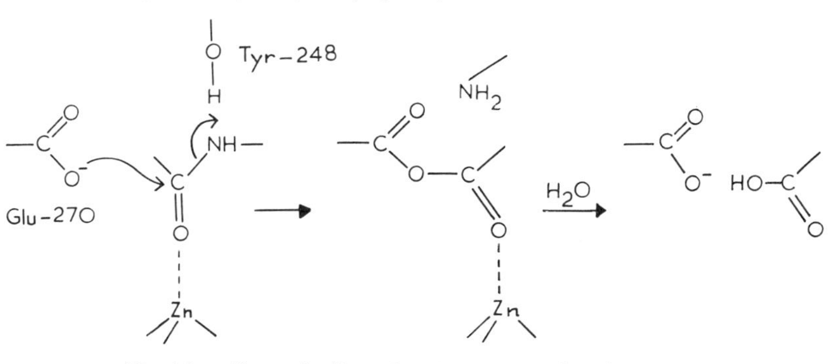

Fig. 4.3 – Alternative Zn-carbonyl mechanism for CPA.

In addition to the above 'zinc carbonyl mechanism' in which the carbonyl group of the peptide is bonded to zinc, a 'zinc hydroxide' mechanism has also been considered, in which a zinc-bound hydroxide ion attacks the carbonyl group of the substrate molecule.

The amino-acid residues near the active site of CPB are shown in Fig. 4.4.

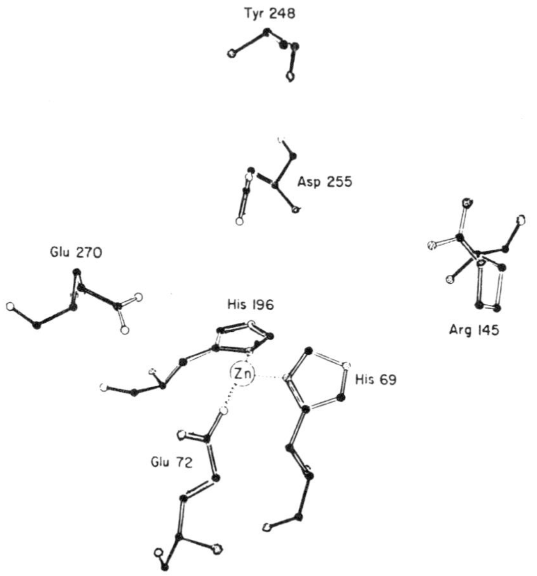

Fig. 4.4 – Amino-acid residues near to the active site of CPB. [Reproduced by permission from *J. Mol. Biol.*, **103**, 175 (1976)].

Aspartic acid-255 lies at the rear of a hydrophobic pocket, which presumably binds the basic side chain of a preferred substrate. In CPA, Asp-255 is replaced by the neutral isolucine which may account for the difference in substrate specificity between CPA and CPB. Chemical modification studies indicate that Glu-270 and Tyr-248 are involved in the catalytic mechanism whereas Arg-145 which is believed to be protonated helps to bind the terminal CO_2^- group of the peptide, thus accounting for the specificity of cleavage of the C-terminal peptide bond.

One obtains only a static picture of an enzyme or an enzyme-substrate complex from structural determinations of the type discussed above, and it can be dangerous to draw conclusions regarding reaction dynamics from such a static picture alone.

MODELS FOR CARBOXYPEPTIDASE

A number of model systems relevant to the action of CPA have been devised. Some of these have involved the use of kinetically inert cobalt(III) complexes, to allow mechanistic features to be defined without complications from ligand exchange processes. A variety of studies have demonstrated the ability of cobalt(III) complexes to promote the hydrolysis of amide, ester and peptide ligands. Two pathways (A and B in Fig. 4.5) for hydrolysis are observed. In

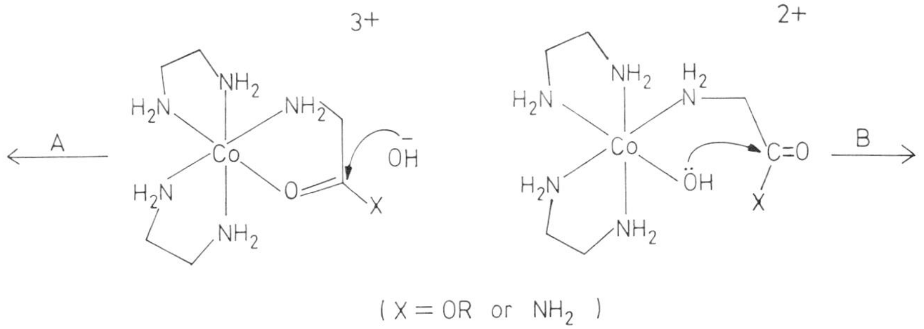

Fig. 4.5 – Pathways for hydrolysis of amide and ester ligands in Co(III) complexes.

mechanism A a cobalt-polarized carbonyl group is attacked by external hydroxide ('bound substrate-free nucleophile'), whereas in mechanism B a free non-coordinated carbonyl group is attacked by cobalt-bound hydroxide ion ('bound nucleophile-bound substrate') (pK_a $Co-OH_2$ is *ca* 6). Both processes are substantially faster than bimolecular hydrolysis of the uncoordinated amides, esters and peptides and the hydrolysis of unidentate complexes of the type $[Co(NH_3)_5(NH_2CH_2CONH_2)]^{3+}$. Pathway B is normally substantially faster than pathway A. At pH 9, the rate enhancement for hydrolysis of glycine amide via pathway B is $\geqslant 10^7$ over that of pathway A, which is $\geqslant 10^{11}$ that for base hydrolysis of uncoordinated glycinamide.

Buckingham and Sargeson's [8a] results establish, that in a model system hydrolysis of a coordinated amide or ester by intramolecular attack of metal hydroxide is far more efficient than intermolecular attack by solvent OH^- on metal-bonded carbonyl. Although the basicity of OH^- is substantially reduced on coordination to cobalt(III), it is a much more effective nucleophile than solvent OH^- when it can take part in an intramolecular reaction.

Substantial rate accelerations have also been observed using labile metal complexes. Thus the anhydride (**4.2**) hydrolyses at a rate independent of pH in the range 1-7.5. ($k_{obs} = 2.7 \times 10^{-3}\ s^{-1}$). The zinc(II) complex, believed to have the structure (**4.3**) hydrolyses at pH 7.5 in a reaction which is first order in the

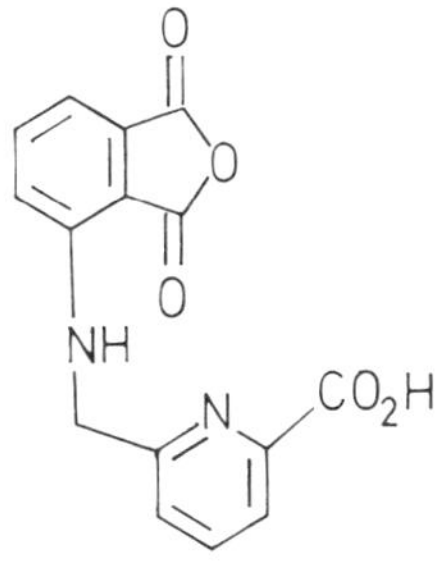

(4.2)

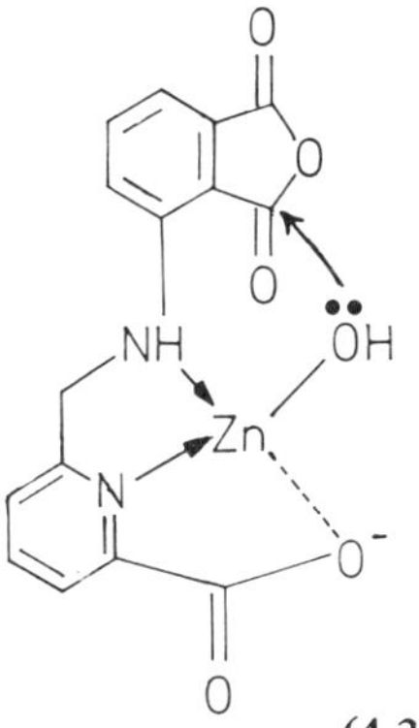

(4.3)

complex and first order in hydroxide ion with $k_{obs} = 3\ s^{-1}$ at 25 °C. The reaction may involve attack by coordinated hydroxide on the anhydride carbonyl. The hydrolysis of 8-acetoxyquinoline-2-carboxylic acid (HA), **(4.4)** follows the rate law, rate = $k_0\,[A^-] + k_{OH}\,[A^-]\,[OH^-]$ (where A^- is the anion)

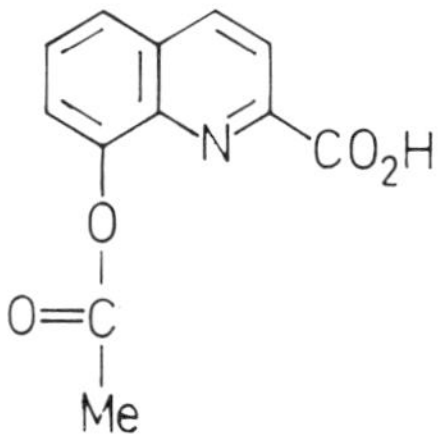

(4.4)

with $k_0 = 1.67 \times 10^{-4}\ s^{-1}$ and $k_{OH} = 0.84\ M^{-1}\ s^{-1}$ at 25 °C. The 1:1 metal complexes of the ester MA^+ (M^{2+} = Zn(II) and Cu(II)) undergo base hydrolysis 2×10^8 times faster than A^-. The metal-promoted reactions involve intramolecular attack by coordinated hydroxide (Fig. 4.6), and this effect in conjugation with a perturbation of 6 pK_a units in the acidity of the conjugate acid of the leaving group leads to rates of hydrolysis ($k = 5 \times 10^1$ to $5 \times 10^3\ s^{-1}$) comparable to the reported values of k_{cat} for the hydrolysis of a good ester substrate by carboxypeptidase A (k_{cat} *ca* $2.3 \times 10^2\ s^{-1}$).

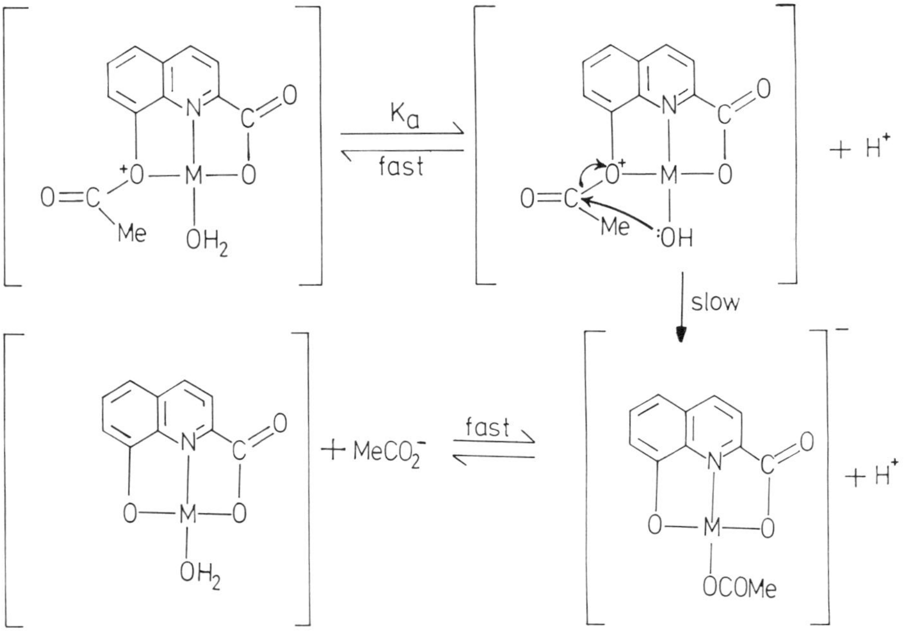

Fig. 4.6 – Mechanism for the metal-promoted reactions (tetrahedral intermediates are assumed).

CARBONIC ANHYDRASE

Carbonic anhydrase is a zinc metalloenzyme present in animals, plants and certain microorganisms which catalyses the reversible hydration of carbon dioxide (equation 4.1). In addition the enzyme also catalyses the hydration of many

$$CO_2 + H_2O \rightleftharpoons HCO_3^- + H^+ \qquad (4.1)$$

aldehydes (equation 4.2) and the hydrolysis of various esters (e.g. *p*-nitrophenyl-

$$CH_3CHO + H_2O \rightleftharpoons CH_3CH(OH)_2 \qquad (4.2)$$

acetate) and sultones. A number of slightly different enzymes, carbonic anhydrases A, B and C occur in different organisms. The most well characterized enzymes are the bovine and human carbonic anhydrases B, which are monomeric and contain 1 atom of tightly bound zinc per 30,000 molecular weight.

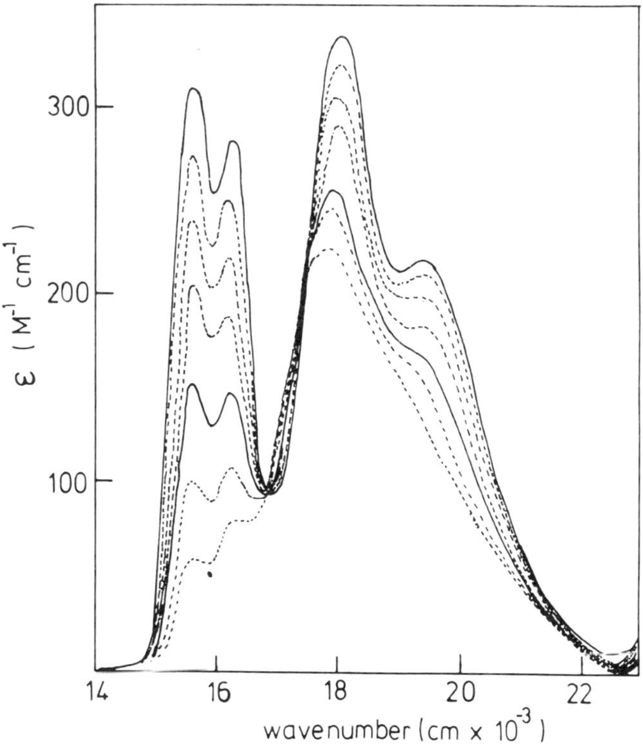

Fig. 4.7 – Electronic spectra of CoHCAB as a function of pH in 10^{-2} M HEPES solutions at pH 6.1, 6.6, 7.1, 7.8, 8.3, 8.6, 9.5 in order of increasing ϵ_{640}. The full lines represent the spectra obtained at the middle and at the end of the titration. From I. Bertini, C; Luchinat and A. Scozzafava, *Inorg. Chim. Acta*, **46**, 85 (1980), reproduced with permission.

The turnover number for CO_2 hydration is *ca* 10^6 s^{-1}, that is 1 mole of the enzyme can hydrate 10^6 moles of CO_2 per second at 37 °C. The uncatalysed hydration rate is 7×10^{-4} s^{-1}, so that the rate acceleration with carbonic anhydrase is of the order of 10^9.

Lindskog first succeeded in removing the very tightly bound zinc under conditions mild enough to avoid denaturation. The Zn was complexed by *o*-phenanthroline and the resulting complex removed by dialysis. The metal-free protein (apoenzyme) was completely inactive, and the activity could be restored by the addition of zinc(II) in the molar ratio of 1 mole of zinc per mole of apoenzyme. Optical rotatory dispersion studies have shown that the apoenzyme and the native enzyme have the same gross tertiary structure, confirming that the function of the zinc is not to stabilize the tertiary structure, but is directly involved in the catalytic activity of the enzyme.

By reacting the apoenzyme with various metal ions it is possible to prepare various other metallocarbonic anhydrases and the Cu(II), Co(II) and Co(III) enzymes have been prepared. The cobalt(II) enzyme shows some catalytic activity and the d–d spectrum of the cobalt(II) derivative which is high spin is markedly pH-dependent, Fig. 4.7. The plot of the molar absorbance at 640 nm against pH does not follow a simple pattern as expected for a single ionizing group, but is consistent with two ionizing groups with pK_a values <6 and >7.

X-Ray crystallographic studies of the human enzyme have established that the molecule is roughly ellipsoidal ($40 \times 45 \times 55$ Å) with the zinc(II) lying near the bottom of a deep cleft near the centre of the molecule. The zinc ion is ligated by three histidine residues (His-117, His-93 and His-95) in a distorted tetrahedral geometry, with the fourth coordination site presumably occupied by a water molecule.

Imidazole is a competitive inhibitor of the hydration of CO_2, and the crystal structure of the human carbonic anhydrase B–imidazole complex has recently been determined. The imidazole is bound weakly to the zinc in the active site, being located in a possible fifth coordination site without displacing the water molecule. This result is consistent with previous observations that imidazole binding produces characteristic changes in the d–d spectrum of the cobalt(II) enzyme. Detailed spectroscopic studies on cobalt(II) bovine carbonic anhydrase alone, and in the presence of inhibitors, have lead to the observation of a previously unrecorded weak absorption band ($\epsilon \sim 10$) at 13,500 cm^{-1} (740 nm) in the presence of halide, acetate and benzoate, but not in their absence. This band is considered to be indicative of the presence of a five-coordinate species.

The mechanism of action of carbonic anhydrase is the subject of considerable controversy. The various mechanisms proposed assume either ionization of a histidine group (bound or not to the zinc) and nucleophilic attack on CO_2 by the coordinated imidazolate ion (Fig. 4.8), or ionization of the Zn(II)-coordinated water and nucleophilic attack by OH^- on CO_2 (Fig. 4.9).

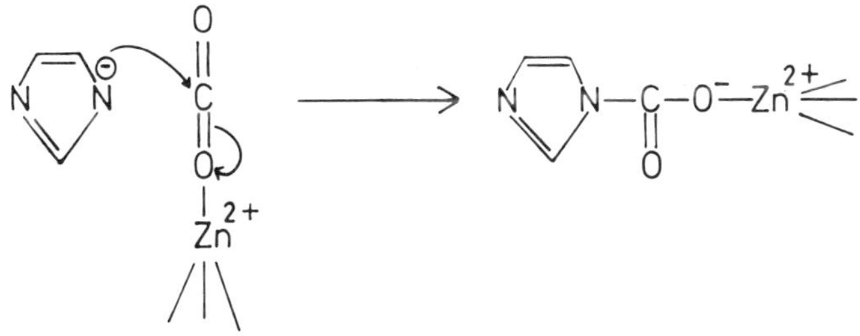

Fig. 4.8 – 'Zinc carbonyl' mechanism.

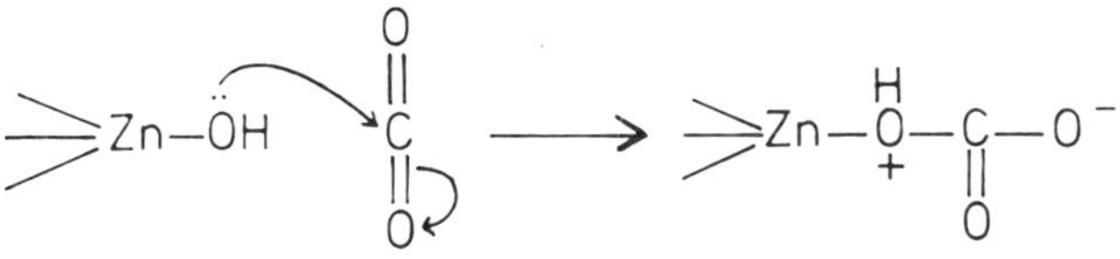

Fig. 4.9 – 'Zinc hydroxide' mechanism.

Kannen and his coworkers [12, 13] as a result of X-ray investigations on the human carbonic anhydrase B-imidazole inhibitor complex have proposed the hydration-dehydration mechanism shown in Fig. 4.10. Glutamic acid-106

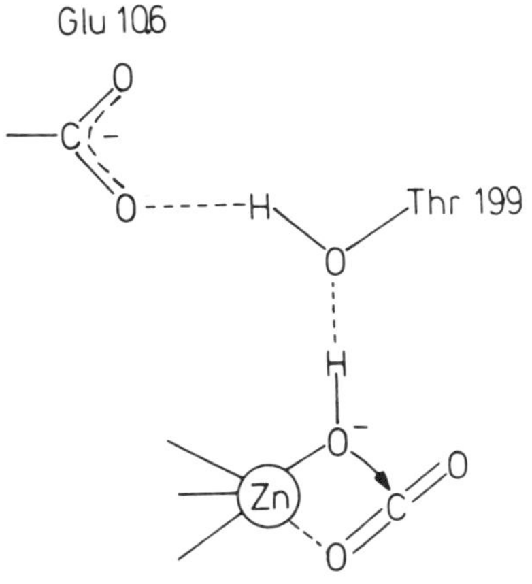

Fig. 4.10 – Proposal for the hydration–dehydration mechanism of carbonic anhydrase. The pK of the zinc bound water is envisaged to be lowered to about 7.0 by the charge distribution on the metal ion and also helped by Glu-106 by the hydrogen bonding through Thr-199.

acts as a general base, and the zinc(II) is a five-coordinate with the zinc providing some Lewis acid catalysis as CO_2 is considered to be bonded to zinc in the fifth site. The mechanism incorporates features of both the 'zinc hydroxide' and zinc carbonyl mechanisms. The ready availability of apocarbonic anhydrase has led to many interesting studies dealing with the binding of metal ions to the apoenzyme. The formation constant for cobalt(II)-carbonic anhydrase at pH 5.0 has been determined kinetically to be log $K = 5.8 \pm 1$ M^{-1} which compares well with the value 6.0 ± 3 M^{-1} calculated from equilibrium dialysis measurements. In the later measurements the formation constant is calculated from the final equilibrium concentrations of the metalloprotein and the metal complex of the competing ligand (e.g. 1,10-phananthroline), the free metal concentration and the known formation constants of the metal(II)-*o*-phenanthroline complex.

It is usually necessary to determine the formation constants for metalloproteins at a single pH, because of possible conformational changes in the protein as the pH varies. Knowledge of the state of protonation of the ligand donor atoms is generally not known, and thus the constants are not pH independent.

The kinetics of formation of the metalloenzyme from Zn(II) and the apoenzyme have also been studied. The reaction is first order in Zn(II) and first order in apoenzyme with $k = 10^4$ M^{-1} s^{-1} at 25 °C. Although the reaction of zinc(II) with the protein is rapid, the rate constant is at least 10^2 lower than for the reaction of zinc(II) with small chelating ligands where rate constants of 10^6 to 10^8 M^{-1} s^{-1} are commonly observed. The enthalpy of activation for the formation of the metalloprotein is about 86 kJ mol^{-1}, this substantial value being compensated by a large positive entropy of activation ($\Delta S^{\ddagger} = 502$ JK^{-1} mol^{-1}. The corresponding thermodynamic parameters for the reaction of small ligands with zinc(II) are $\Delta H^{\ddagger} = 26 - 30$ kJ mol^{-1}. By using the formation of zinc(II)-carbonic anhydrase, it is possible to calculate the dissociation rate constant (k_d, eq. (4.3)) from

$$\text{apoenzyme} + \text{Zn(II)} \underset{k_d}{\overset{k_f}{\rightleftharpoons}} \text{carbonic anhydrase} \tag{4.3}$$

the relationship $K = k_f/k_d$. The value of k_d is about 1.5×10^{-9} s^{-1} at 25 °C.

MODELS FOR CARBONIC ANHYDRASE

Inorganic chemists have studied a variety of metal complexes as possible models for carbonic anhydrase. Considerable attention has been devoted to metal complexes with coordinated hydroxide as a possible nucleophile ('zinc hydroxide' mechanism). Thus the metal complex (**4.5**) $[ZnCROH]^{+}$, which is five-coordinate, is an effective catalyst for the hydration of acetaldehyde. Table 4.4 lists rate constants for the hydration of CO_2 and the hydrolysis of *p*-nitrophenyl acetate by various species. Simple hydroxo complexes do display some catalytic activity and the activity parallels the basicity of the species as determined by the pK_a of the conjugate acid. The enzyme is of course vastly more active than would be expected on the basis of its pK_a, due to the cooperative effect of a variety of catalytic groups at the active site.

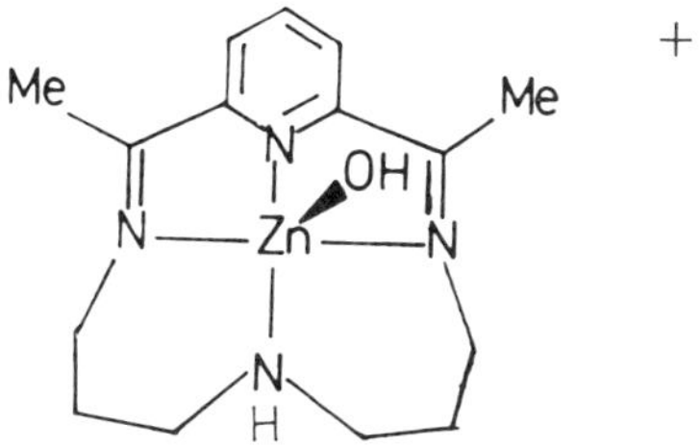

(4.5)

Table 4.4 – Rate constants for the hydration of CO_2 and the hydrolysis of *p*-nitrophenyl acetate.

	$k/M^{-1}\ s^{-1}$, 25 °C	pK_a
(a) CO_2 hydration		
OH^-	8500	15.5
Carbonic anhydrase	10^7–10^8	8
$[(H_3N)_5CoOH]^{2+}$	220	6.4
H_2O	6.7×10^{-4}	−1.7
$[(NH_3)_5CrOH]^{2+}$	10	5.2
$[(NH_3)_5RhOH)]^{2+}$	470	6.78
$[(NH_3)_5 IrOH]^{2+}$	590	6.70
(b) *p*-Nitrophenyl acetate hydrolysis		
OH^-	9.5	15.5
Imidazole	0.58	14
$[(NH_3)_5 Co Im]^{2+}$†	9	10.02
Carbonic anhydrase	460	7.5
$[(NH_3)_5 CoOH)]^{2+}$	1.52×10^{-3}	6.4

†Contains deprotonated imidazole. The pK_a values relate to the conjugate acids.

One important aim of this type of study is to determine if a postulated mechanism is chemically reasonable. In this case it is clear that hydroxide ion bound to zinc(II) is capable of catalysing the hydration of CO_2 and the hydrolysis of esters which are substrates for the enzyme.

ALCOHOL DEHYDROGENASES

These enzymes catalyse the reversible oxidation of alcohols to aldehydes and are pyridine-nucleotide dependent enzymes. The horse liver enzyme has been the subject of much detailed study. The enzyme consists of two identical subunits of $M = 40{,}000$ which contains 4 gram atoms of zinc per mole and binds 2 moles of NAD^+ per mole of enzyme. X-Ray diffraction studies indicate that there are two types of zinc ion. The active site zinc is situated at the bottom of a hydrophobic pocket in a deep cleft about 25 Å from the protein surface. The ligands on zinc are two cysteinyl sulphurs (Cys-46 and Cys-174) and an imidazole group of histidine (His-67). The distorted tetrahedral stereochemistry is completed by a water molecule. The second type of zinc is close to the surface of the enzyme, but not exposed to solvent. In this case the ligand groups are four cysteinyl sulphurs (Cys-97, -100, -103 and -111). The active site of zinc is considered to act as an electrophilic centre, promoting hydride transfer to NAD^+ via formation of a zinc(II) alkoxide complex. The overall reaction can be written as shown in Fig. 4.11. The function of the second zinc centre is unknown, this region of the enzyme has been called a 'molecular fossil' as it looks like a catalytic centre reminiscent of the FeS_4 proteins.

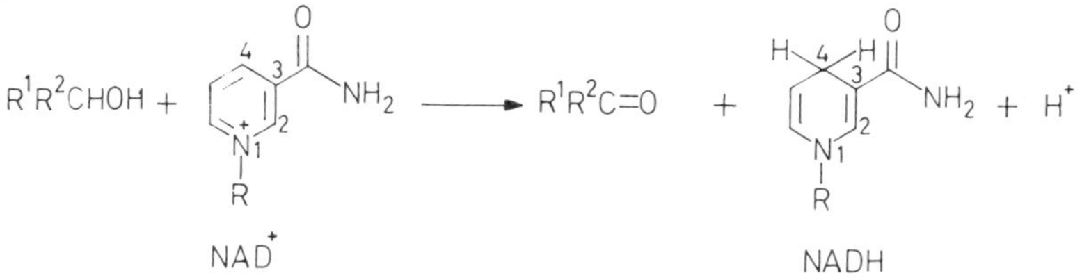

Fig. 4.11

The sequence of events during the oxidation of an alcohol to a carbonyl compound by alcohol dehydrogenases is considered to be as follows. The enzyme binds NAD^+ which induces a conformational change and causes one proton to be released. ($Zn\text{--}OH_2 \rightarrow Zn\text{—}OH$); the alcohol substrate binds near the active zinc and is converted to an alkoxide ion. Loss of H_s from C-1 of the alkoxide to the *re* face (C-4) of the pyridinium ring of NAD^+ produces aldehyde (or ketone) which dissociates from the enzyme. After a conformational change NADH is released. A sketch of the active site is shown in Fig. 4.12.

A variety of roles for zinc(II) in the above scheme have been proposed. It has been usual to consider the reaction in the direction aldehyde (or ketone) being converted to the appropriate alcohol. These various roles may be summarized as follows:

(i) Replacement of zinc-bound water by the substrate occurs, leading to polarization of the substrate carbonyl group by zinc.

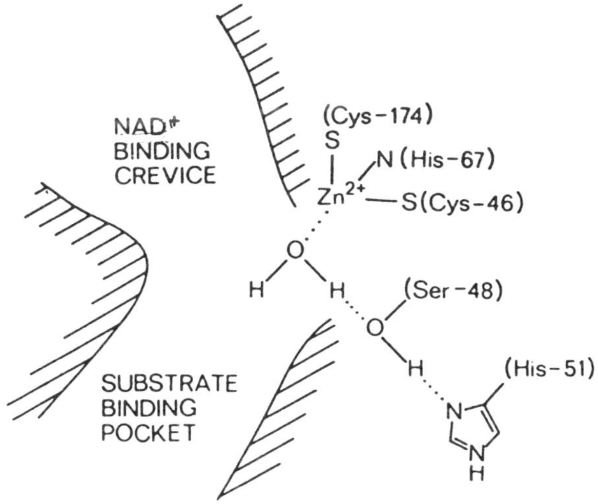

Fig. 4.12 – Sketch of the active site of horse liver alcohol dehydrogenase (courtesy of C-I Brändén).

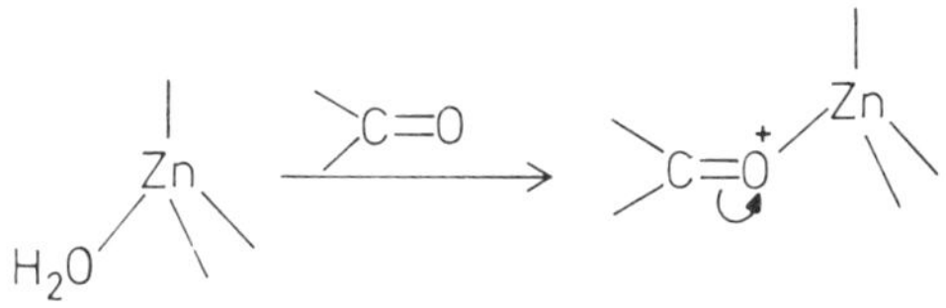

(ii) Zinc-bound water activates a substrate molecule by hydrogen-bonding.

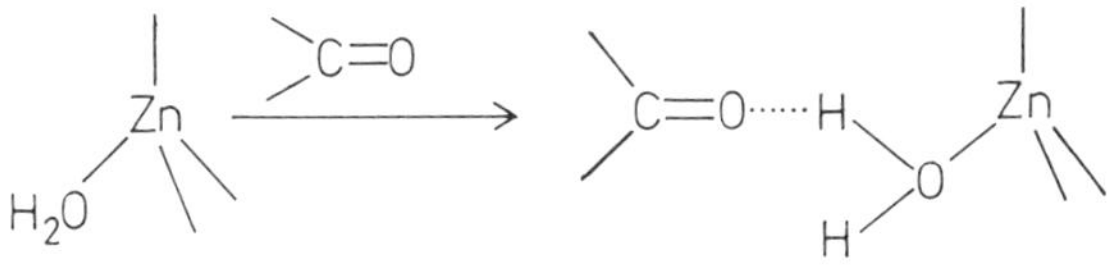

(iii) There is an intermediate five-coordinate zinc(II) complex containing a substrate molecule and water (cf. carbonic anhydrase). The substrate molecule is activated by coordination to Zn and by proton transfer from the zinc-bound water which is coupled to Ser-48 and His-51 (Fig. 4.13).

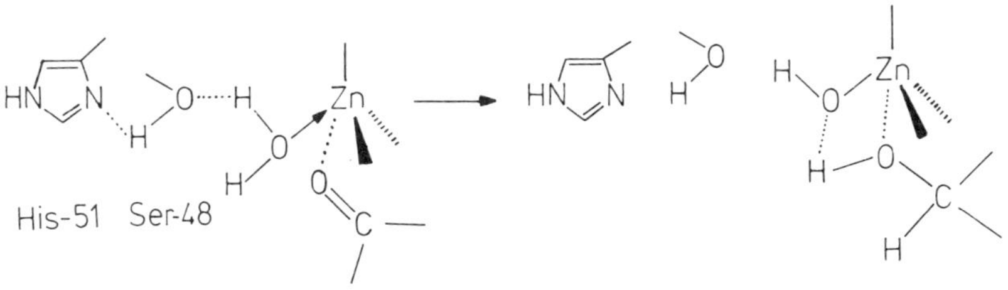

Fig. 4.13

The zinc functions primarily as a Lewis acid in the three postulated mechanisms, enhancing the susceptibility of the carbonyl group to nucleophilic attack by hydride ion or electron transfer from NADH.

Support for all three mechanisms has come from a variety of studies. A number of model systems have been investigated to study the influence of Zn(II) and other metal ions on the reactivity of C=O and CHOH groups. It has been shown, for example, that Zn(II) catalyses the reduction of 1,10-phenanthroline-2-carboxyaldehyde (**4.6**) to the corresponding carbinol by the 1,4-dehydronicotinamide (**4.7**) in acetonitrile solution.

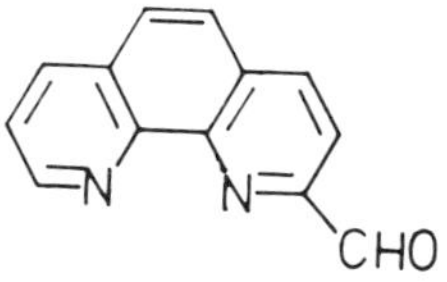

(**4.6**)

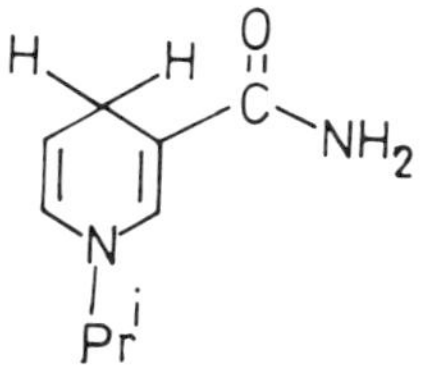

(**4.7**)

The reduction does not occur in the absence of zinc(II) over a period of 4 days. Presumably the reaction involves the formation of the zinc(II) complex (**4.8**) and the sequence of reactions shown in Fig. 4.14.

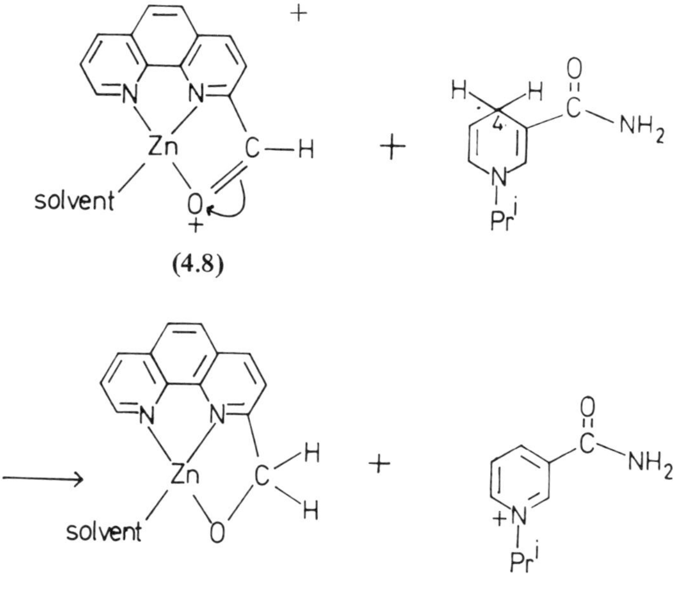

(**4.8**)

Fig. 4.14 – Reduction of 1,10-phenanthroline-2-carboxyaldehyde.

Reduction is slower if the two hydrogen atoms at C-4 are replaced by deuterium (k_H/k_D = 1.74) confirming that cleavage of the C_4–H bond occurs in the rate determining step.

It will be recognized from the above examples that in spite of the availability of very detailed structural information obtained from X-ray work, the mechanistic details of the action of some metalloenzymes remains obscure and controversial. It is also apparent that studies of simple 'model' metal complexes will continue to play an important role in helping to define plausible schemes for the action or metalloenzymes.

REFERENCES AND BIBLIOGRAPHY

Metalloenzymes

[1] R. J. P. Williams and A. E. Dennard. 'Transition Metal Ions as Reagents in Metalloenzymes, *Transition Metal Chemistry,* ed. R. L. Carlin, Edward Arnold, London, 1966, Vol. 2.

[2] J. E. Coleman, 'Metal Ions in Enzymatic Catalysis' in *Progress in Bioinorganic Chemistry,* ed. E. T. Kaiser and F. J. Kezdy, Vol. 1, Wiley-Interscience, New York, 1971.

[3] M. L. Bender, *Mechanisms of Homogeneous Catalysis from Protons to Proteins,* Wiley-Interscience, New York, 1971.

[4] A. R. Fersht, *Enzyme Structure and Mechanism,* Freeman, Reading, 1977.

Metal Ion Catalysis

[5] R. W. Hay, *Metal Ions in Biological Systems,* Vol. 5, ed. H. Sigel, Marcel Dekker, New York, 1976. Deals with catalysis of ester hydrolysis and decarboxylation.

[6] M. M. Jones, *Ligand Reactivity and Catalysis,* Academic Press, New York, 1968. A general discussion of catalysis and the reactions of coordinated ligands.

[7] D. P. N. Satchell, 'Metal-ion-promoted Reactions of Organo-sulphur Compounds, *Chem. Soc. Rev.,* **6**, 345 (1977).

[8] R. W. Hay, 'Metal Ion Catalysis and Metalloenzymes, *Chapter 4, in An Introduction to Bioinorganic Chemistry,* ed. D. R. Williams, C. C. Thomas, Springfield, Illinois, 1976.

[8a] N. E. Dixon and A. M. Sargeson, 'Roles for the Metal Ion in Reactions of Coordinated Substrates and in Some Metalloenzymes; in *Zinc Enzymes,* ed. T. G. Spiro, John Wiley and Sons, New York, 1983.

[8b] D. P. N. Satchell and R. S. Satchell, 'Kinetic Studies of Metal Ion Catalysis of Heterolytic Reactions', *Ann. Rep. Chem. Soc. A,* 25, (1978).

Specific Enzymes

Carbonic Anhydrase

[9] S. Lindskog, L. E. Henderson, K. K. Kannen, A. Liljas, P. O. Nyman and B. Strandberg, in *The Enzymes,* ed. P. D. Boyer, 3rd edn., p. 587, Academic Press, New York.

[10] J. E. Colman, *Prog. Bioorg. Chem.,* **1**, 159 (1971).

[11] A. Liljas, K. K. Kannen, P.-C. Bergsten, I. Waara, K. Fridborg, B. Strandberg, U. Carlborn, L. Järup, S. Lövgren and M. Petef. *Nature New Biology, Lond.,* **235**, *131 (1972). X-Ray data.*

[12] K. K. Kannen, B. Nostrand, K. Fridborg, S. Lovgren, Petef, *Proc. Nat. Acad. Sci. USA,* **72**, 51 (1975). X-Ray data).

[13] K. K. Kannen, M. Petef, K. Fridborg, H. Cid-Dresdner and S. Lovgren, *FEBS Letters,* **73**, 115 (1977). Structure and function of carbonic anhydrases and the X-ray structure of the imidazole inhibitor complex.

[14] R. W. Hay, *Inorg. Chim. Acta,* **46**, L115 (1980).

Carboxypeptidase

[15] W. N. Lipscombe, *Chem. Soc. Rev.,* 1972, **1**, 319.

[16] W. N. Lipscombe, *Tetrahedron,* 1974, **30**, 1725.

[17] E. T. Kaiser and B. L. Kaiser, *Acc. Chem. Res.,* **5**, 219 (1972).

[18] R. W. Hay and P. J. Morris, 'Metal Ion Promoted Hydrolysis of Amino Acid Esters and Peptides', in *Metal Ions in Biological Systems,* Vol. 5, ed. H. Sigel, Marcel Dekker, New York, New York, 1976.

[19] D. A. Buckingham, 'Metal-OH and its Ability to Hydrolyse (or Hydrate) Substrates of Biological Interest', in *Biological Aspects of Inorganic Chemistry,* eds. A. W. Addison, W. R. Cullen, D. Dolphin and B. R. James, Wiley-Interscience, New York, New York, 1977.

Alcohol Dehydrogenase

[20] C. I. Bröndén, H. Jornvall, H. Eklund and B. Furugren, *The Enzymes,* **11**, 104 (1975).

[21] R. T. Dworschack and B. V. Plapp, *Biochemistry,* **16**, 2716 (1977).

[22] D. J. Creighton, J. Hajdu and D. S. Sigman, *J. Am. Chem. Soc.,* **98**, 4619 (1976). Model system.

Enzymes

For a general discussion of enzymes and enzyme catalysis see *Enzyme Structure and Mechanism* by A. R. Fersht, W. H. Freeman, Reading, 1977. Those requiring a detailed discussion of the background to enzymic catalysis should consult *Catalysis in Chemistry and Enzymology*, by W. P. Jencks, McGraw-Hill, New York, 1969.

CHAPTER 5

Oxygen Carriers and Oxygen Transport Proteins

INTRODUCTION

The solubility of dioxygen (O_2) in water is quite low. Thus 1 dm^3 of water dissolves 6.59 cm^3 of O_2 at 1 atmosphere pressure and 20 °C giving a 3×10^{-4} M solution. The rate at which O_2 can be delivered by the circulatory system is therefore limited. As a result biological systems have developed O_2 carriers by reversibly coordinating O_2 to a transition metal (Fe, Cu or possibly V) bound to a protein. Hence human blood dissolves 200 cm^3 of O_2 per dm^3 when in equilibrium with air at 20 °C giving a 9×10^{-3} M solution. Blood carries 30 times as much oxygen as pure water.

A variety of respiratory pigments occur (haemoglobin, haemerythrin and haemocyanin) and these are briefly described in the following sections. Some properties of the oxygen binding pigments are summarized in Table 5.1.

Table 5.1 – Comparison of some properties of the oxygen binding pigments.

	Haemoglobin	Haemerythyrin	Haemocyanin
Metal	Fe	Fe	Cu
Metal: O_2	Fe : O_2	2Fe : O_2	2Cu : O_2
Oxidation state of metal in the deoxy protein	Fe(II)	Fe(II)	Cu(I)
Coordination of metal	Porphyrin	Protein side chains	Protein side chains
No. subunits	4	8	Variable
Mol. wt.	65,000	108,000	400,000–9,000,000
Colour			
Oxygenated	Red	Violet–pink	Blue
Deoxygenated	Red–blue	Colourless	Colourless

From D. M. Kurtz, D. F. Shriver and I. M. Klotz, *Coord. Chem. Rev.*, **24**, 145–178 (1977).

IRON PORPHYRINS (HAEMOGLOBINS AND MYOGLOBINS)

The interaction of molecular oxygen with haemeproteins is of primary importance in respiratory and metabolic processes. These haemeproteins bind one molecule of O_2 per Fe(II).

Only two different porphyrins have been implicated in these reactions, protoporphyrin IX and chlorocruoroporphyrin. The structure of protoporphyrin IX is shown in (**5.1**) and its Fe(II) complex in (**5.2**). The iron(II) complex is often known as haem *b*, and formation of the complex involves deprotonation of the two N-H groups of the porphyrin ring. Chlorocruoroporphyrin has a formyl group (-CHO) in place of the vinyl group at position 2 of structure (**5.1**).

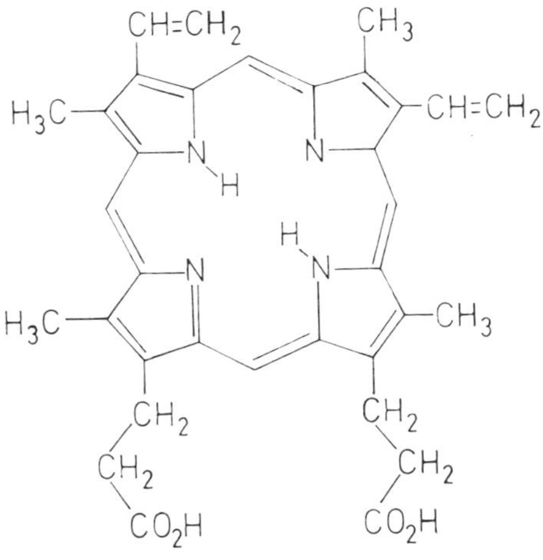

(5.1)

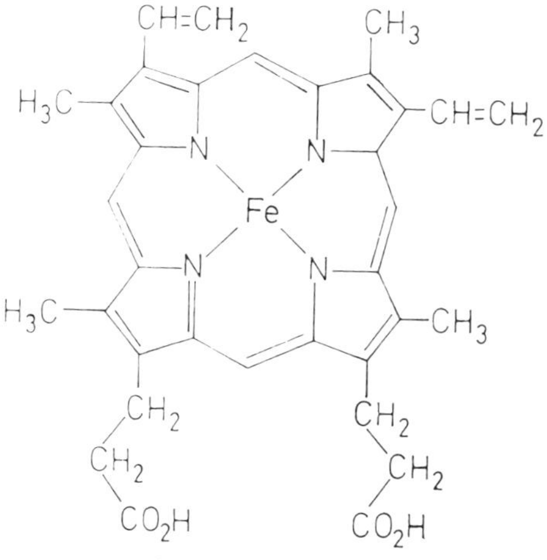

(5.2) Iron(II) protoporphyrin (IX) (haem *b*)

Combinations of proteins and iron(II) chlorocruoroporphyrin are called chlorocruorins and have only limited occurrence. Proteins (globins) and iron(II) protoporphyrin IX are very widespread and occur in almost all vertebrates. They are present in blood, when they are called *haemoglobins* and in muscle (*myoglobins*). These compounds are abbreviated to Hb (haemoglobin) and Mb (myoglobin).

Myoglobin (Mb) the oxygen-binding protein in muscle tissue is responsible for the storage and possible transport of oxygen across membranes, and consists of a 153 residue peptide and iron(II) protoporphyrin IX. The iron(II) protoporphyrin IX is held within a cleft in the protein principally by non-covalent, largely apolar interactions. The only covalent linkage between porphyrin and protein arises from a coordinate bond between the so-called proximal imidazole of histidine residue F-8 and the Fe(II). In the deoxy (i.e. oxygen free) form of Mb the Fe(II) is five-coordinate, high spin, with an ionic radius too large to fit within the cavity of the porphyrin. Consequently the high spin iron atom projects out of the mean plane of the four porphyrin nitrogens towards the proximal imidazole and away from the oxygen binding site. This situation is illustrated in (**5.3**) and (**5.4**).

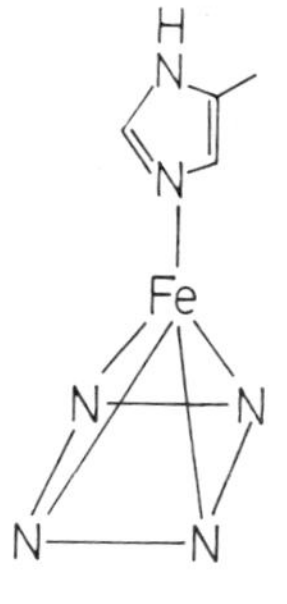

(5.3)

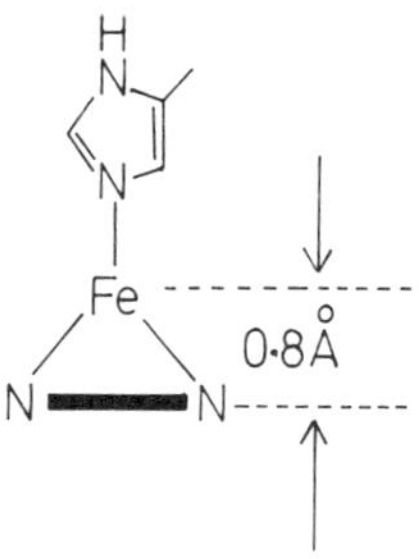

(5.4)

The low spin iron atom of diamagnetic oxymyglobin is thought to lie in the porphyrin plane. On reacting with O_2 the iron atom moves into the plane of the porphyrin ring pulling along the proximal imidazole. This movement of the proximal imidazole may be transferred through the four Mb like subunits of haemoglobin contributing to the cooperativity of oxygen binding between the four distant iron porphyrins in Hb.

The active site in Hb and Mb is the haem which is tightly bound to a protein (globin) through about 80 hydrophobic interactions and a single coordinate bond between imidazole of the 'proximal' histidine and iron(II). Mb is a monomer whereas Hb is a tetramer composed of two similar globins of unequal length. In spite of numerous differences in their amino acid sequences, all Mb and Hb globin-haem units have very similar tertiary structures consisting of eight helical regions designated by A–H. Residues are numbered by their order in these segments. Thus the proximal histidine which is common to all functional Mbs and Hbs, is invariably the eighth residue in the helical region F and hence the nomenclature His-F8. The haem is wedged in a crevice between segments E and F. Oxygen binds on the E ('distal') side of the porphyrin.

The most significant property of Hb is cooperative O_2 binding, the oxygen affinity of the tetramer rises with increasing oxygen saturation. The dioxygen binding curve is autocatalytic ('cooperative') not hyperbolic. The degree of cooperative binding can be expressed by the Hill constant n, which has the value of approximately 2.8. Independent O_2 binding by the four haems of Hb would require $n = 1.0$ whereas all-or-none binding to the four haems would require $n = 4$.

The oxygen saturation curves for myoglobin and haemoglobin are shown in Fig. 5.1. Haemoglobin is less efficient at O_2 uptake under low O_2 pressures where myoglobin is very efficient. In muscle tissue, for example, where P_{O_2} is small,

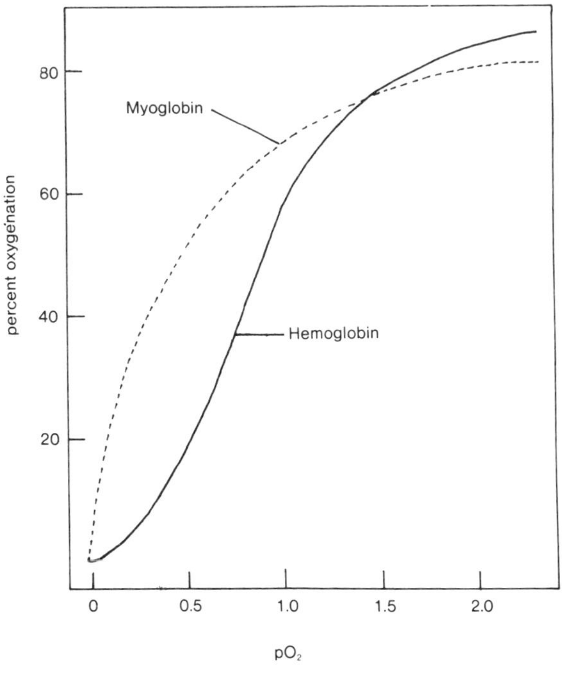

Fig. 5.1 – Oxygen saturation curves of myoglobin and haemoglobin. [From J. M. Rifkind, in *Inorganic Biochemistry*. G. L. Eichhorn (ed.), Vol. 2, Elsevier, New Yor, p. 832 (1973).

there is a thermodynamically favourable O_2 transfer from HbO_2 to MbO_2 in order to pass O_2 into the cell. The oxygenation equilibrium for myoglobin can be represented by equation (5.1).

$$Fe + O_2 \overset{K}{\rightleftharpoons} FeO_2 \text{ with } K = [FeO_2]/[Fe]\,P_{O_2} \tag{5.1}$$

If α is the fraction which is oxygenated, $\alpha = [FeO_2]/\{[Fe] + [FeO_2]\}$ then

$$K = \alpha/(1-\alpha)P_{O_2}$$

$$\alpha = KP_{O_2}/(1 + KP_{O_2})$$

$$\frac{1}{\alpha} = 1 + \frac{1}{K}\frac{1}{P_{O_2}} \tag{5.2}$$

The myoglobin curve in Fig. 5.1 corresponds to equation (5.2), but the haemoglobin curve does not follow such an equation, but an empirically modified form with P_{O_2} replaced by $P^n_{O_2}$ where n is the Hill constant.

In addition, Hb exhibits two other so-called linkage properties; first, the binding of O_2 is pH dependent, although there is no ionizable group in the haem ('Bohr effect'), and second the oxygen affinity is altered by the binding of an organic phosphate to Hb, although the phosphate binding site is far removed from the haems ('phosphate effect').

The deoxy- and oxy-forms of Hb (respectively denoted as T and R) differ both in the conformation of the individual chains (tertiary protein structure) and in the relative orientation of the chains (quaternary protein structure). The linkage effects exhibited by Hb arise from the reversible transition between these two forms.

This phenomenon of 'haem–haem interaction' or cooperativity must have its basis on the molecular level, even though the Fe atoms are separated by ≈ 25 Å. A 'trigger mechanism' has been proposed by Perutz [2] which is based on the ideas of Williams and Hoard. In deoxy-Hb the high spin Fe(II) is about 0.8 Å above the mean plane of the porphyrin ring. Upon oxygenation the iron atom becomes low spin and moves into the plane of the porphyrin ring (5.5 → 5.6). The proximal histidine is also shifted by ≈ 0.8 Å so causing a conformational change in the protein leading to crevice opening.

More recent work with Co(II) in place of Fe(II) (coboglobin, CoHb), however, reveals a similar degree of cooperativity ($n \approx 2.5$) as found in haemoglobin with less than one-half of the histidine displacement. Consequently the problems remain. Recent work by Busch and coworkers [10] on coboglobin model systems indicates that specific steric effects can have marked influences on O_2 binding.

Iron(II) has a d^6 configuration and iron(II) porphyrins exhibit three spin states, each having a characteristic coordination number and structure. Six-coordinate iron(II) haems with two axial ligands are invariably diamagnetic

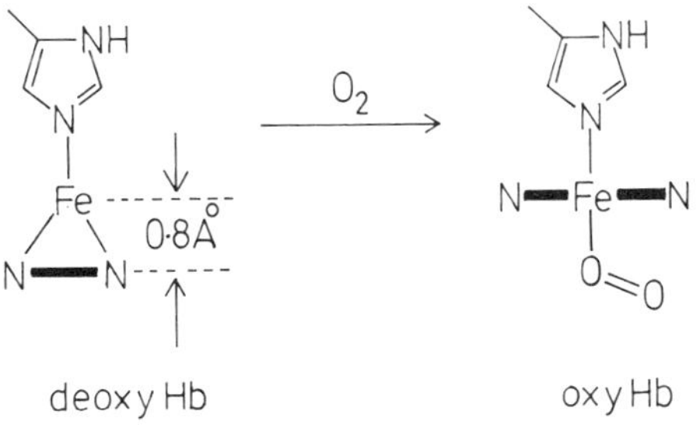

(5.5) (5.6)

and low spin with a $t_{2g}^6\,e_g^0$ configuration ($S = 0$). The covalent radius of low-spin iron(II) is such that it fits into the porphyrin ring without stress, as is observed with MbCO (actually 0.01 Å out of the plane towards CO). High-spin iron(II) porphyrins are invariably five-coordinate with the iron atom displaced well out of the porphyrin ring towards the single axial ligand. The converse that five-coordinate iron(II) porphyrins are invariably high spin is not necessarily true.

Iron(II) porphyrins also exhibit an intermediate spin state ($S = 1$) characterized by FeTPP (TPP = *meso*-tetraphenylporphyrinato). In this state, which has no biological counterpart, iron is four-coordinate and exhibits such short iron-nitrogen distances that the porphyrin develops a pronounced ruffling.

Proposed structural models for dioxygen binding in HbO_2 and MbO_2 include a sideways triangular structure **(5.7)** by Griffith and an end-on angular bond **(5.8)** by Pauling and by Weiss. The former structure would require a formal coordination number of seven for MbO_2 or HbO_2 which would be sterically unfavourable. Although structural analogues of **(5.7)** have been obtained from

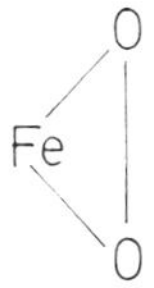

(5.7) **(5.8)**

coordinatively unsaturated d^8 and d^{10} complexes of other transition metals, these are all four- or six-coordinate and none involves a macrocyclic tetradentate ring. Three angular dioxygen complexes of cobalt(II) have been structurally characterized, but these are not isoelectronic with the Fe–O_2 system. Structural studies of the dioxygen complexes of Collman's picket fence porphyrin, which will be discussed in detail in a later section, establishes that O_2 is bound 'end on' with an Fe-O-O bond angle of 136°. These complexes with an axial 1-methylimidazole or 1-*n*-butylimidazole are diamagnetic. The Fe-O distance of 1.75(2) Å is about 0.1 Å shorter than expected from a summation of covalent radii and is also shorter than observed (1.86 Å) in angular O_2 complexes of Co(II). This result may indicate significant multiple bonding. A reasonable description of the nature of the dioxygen in this model system would be coordinated singlet oxygen possibly experiencing modest $d\pi$-$p\pi$ back bonding from iron to O_2. Coordination of low spin d^6 iron(II) by spin singlet O_2 with the configuration $(\pi^*)^2$ $(\pi^*)^0$ occurs, the π^* pair acts as σ-donor to iron(II) while the empty π^* orbital acts as a π-acceptor orbital.

HAEMOCYANINS

Haemocyanins contain Cu and bind one molecule of O_2 for every pair of copper(I) ions. The oxy form is blue, and the deoxy form almost colourless. Haemocyanins are found only in molluscs and arthropods (crustaceans and arachnids), and occur dissolved in the blood where they usually comprise 90–98% of the total protein present. The subunits contain two atoms of Cu and have a molecular weight of 50,000–74,000, they form aggregates with a molecular weight up to 9×10^6.

Currently relatively little is known about the molecular structure of the active site in the haemocyanins. There appears to be at least two imidazole ligands (from histidine residues) per copper. In addition, the reaction stoichiometry 2Cu:$1O_2$, the diamagnetism of the deoxy form, and the resonance Raman spectrum, of oxyhaemocyanin, suggest a Cu(II)-O_2^{2-}-Cu(II) structure for the oxygenated active site.

Inorganic chemists have been interested in developing suitable copper complexes which would mimic some of the properties of haemocyanin (so-called biomimetic chemistry).

Simmons and Wilson [15] have recently described a synthetic copper(I) complex with two imidazole donors which binds dioxygen reversibly in both the solid state, and in solution at room temperature. The preparation of the complex is shown in Fig. 5.2.

Condensation of 2,6-diacetyl pyridine with two moles of histamine give the ligand 'bimp' (2,6-[1-(2-imidazol-4-ylethylimino)ethyl] pyridine. Addition of $[Cu(I)(MeCN)_4](ClO_4)$ under N_2 gives the dark red [Cu(I) (bimp)] (ClO_4). The copper is presumed to be five-coordinate with the complex being monomeric in solution. The use of $Cu(ClO_4)_2.6H_2O$ in the synthesis gives [Cu(II) (bimp)] $(ClO_4)_2$ as a paramagnetic green solid (μ_{eff} at 25 °C = 1.8 BM). If a deoxygenated solution of the red copper(I) complex is exposed to O_2 (1 atm, room temperature) the solution rapidly turns green absorbing 1 mole of O_2 per 2 moles of copper.

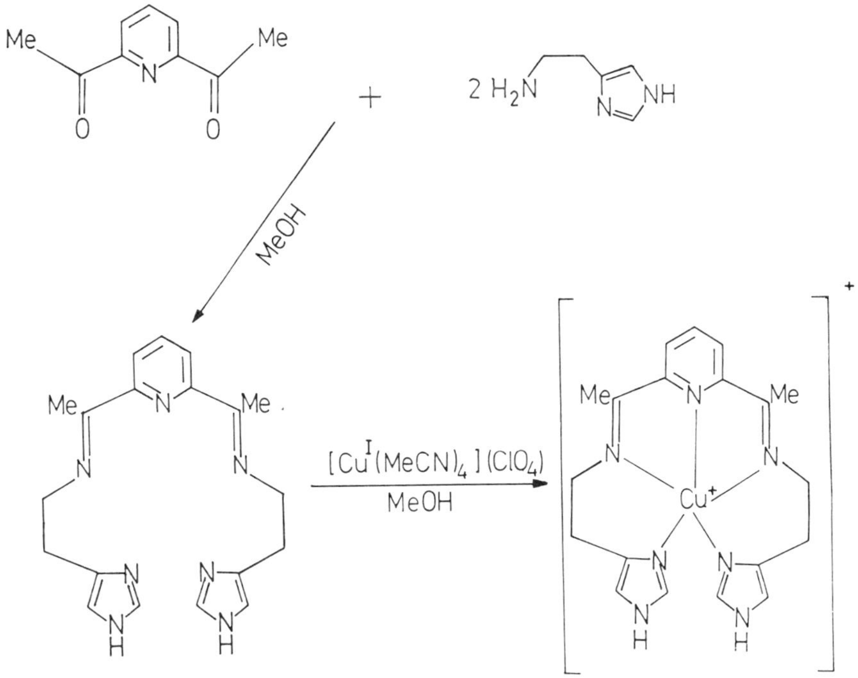

Fig. 5.2

The reaction is complete in *ca* 2 minutes and can readily be reversed by gentle heating (*ca* 40 °C) and degassing with N_2 or by vigorous stirring of the solution under reduced pressure. Under these conditions the original red colour returns and the solution will again absorb O_2. The reaction is believed to involve the equilibria:

$$LCu^{I} + O_2 \rightleftharpoons LCu^{II}O_2^{-}$$

$$LCu^{II}O_2^{-} + LCu^{I} \rightleftharpoons LCu^{II}\text{-}O_2^{2-}\text{-}Cu^{II}L$$

in which the metal is formally oxidized and O_2 formally reduced to $O_2^{\dot{-}}$ or O_2^{2-}. Both adducts might be expected to be diamagnetic via antiferromagnetic exchange between the formally Cu^{II} ($S = \frac{1}{2}$) centres in the binuclear complex or between Cu^{II} ($S = \frac{1}{2}$) and $O_2^{\dot{-}}$ ($S = \frac{1}{2}$) in the mononuclear form. The former situation is usually invoked to account for the diamagnetism and absence of an e.s.r. signal in the oxyhemocyanins.

HAEMERYTHRINS

Haemerythrins contain Fe, but no porphyrin and the Fe is ligated solely by the protein. These iron proteins bind one O_2 per two Fe(II); the oxy form is violet-pink and the deoxy form colourless. The protein from *Golfingia gouldii* has a molecular weight of 108,000 and consists of eight subunits, each of which contain two Fe atoms. The primary structure of the haemerythrin subunit from *P. gouldii* is shown in Fig. 5.3. A schematic drawing of the subunit arrangement of haemerythrin is shown in **(5.9)**.

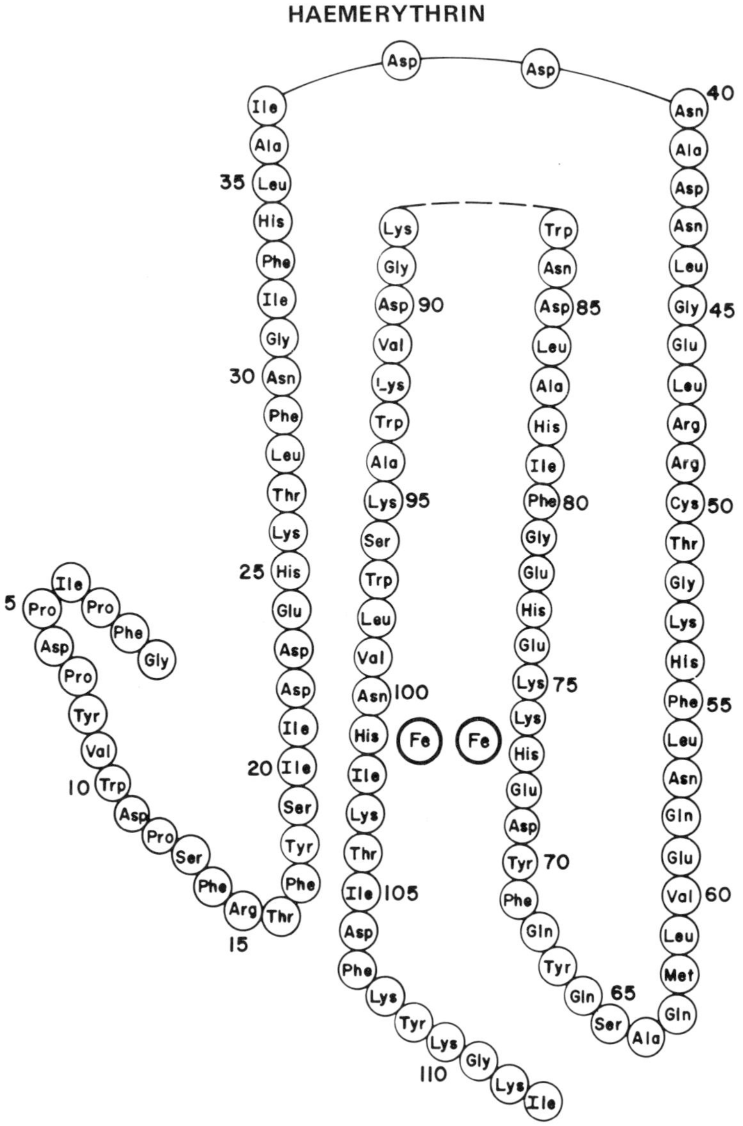

Fig. 5.3 – Primary structure of haemerythrin from erthrocytes of *P. gouldi.* From D. M. Kurtz, Jr., D. F. Schriver and I. M. Klotz, *Coord. Chem. Rev.*, **24**, 145–178/1977). Reproduced by permission of Elsevier.

(5.9)

Schematic drawing of the subunit arrangement of haemerythrin.

Chemical modification of cysteine SH groups with RHg^+ will dissociate the octamer into monomers. Interconversions between octamers and monomers, and between the different chemical states of haemerthyrin are illustrated in Fig. 5.4.

Irreversibly oxidized met- and meso-haemerythrins lose their ability to bind oxygen.

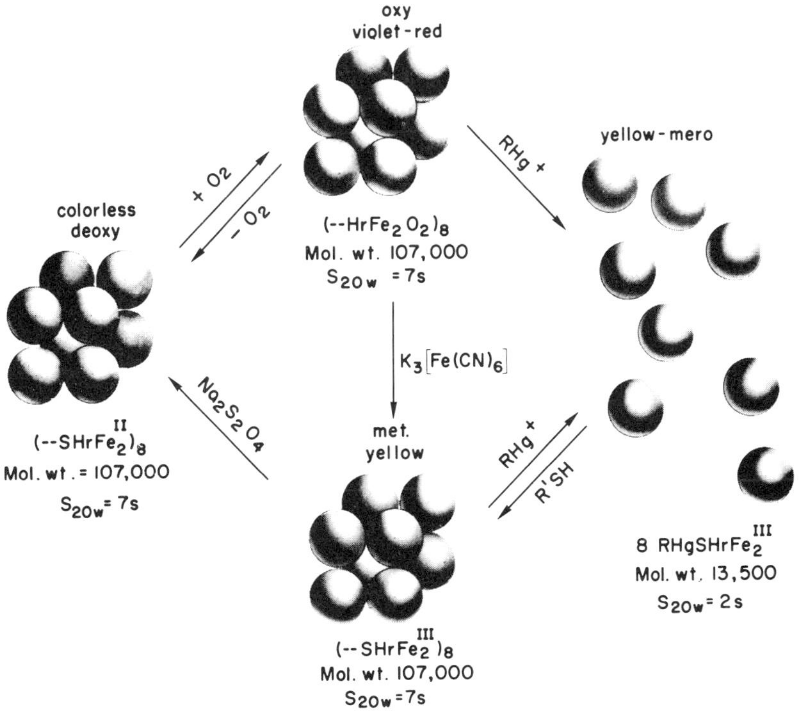

Fig. 5.4 – Macromolecular properties of interrelationships between deoxy-, oxy- and methaemerythrin and their monomeric subunit. From D. M. Kurtz, Jr., D. F. Schriver and I. M. Klotz, *Coord. Chem. Rev.*, **24**, 145–175/1977). Reproduced by permission of Elsevier.

The stoichiometry of the oxygen uptake of 2Fe:1O_2 suggests that the two iron atoms in haemerythrin are in close proximity with the dioxygen, perhaps bridging them. The oxygenation reaction can be formulated

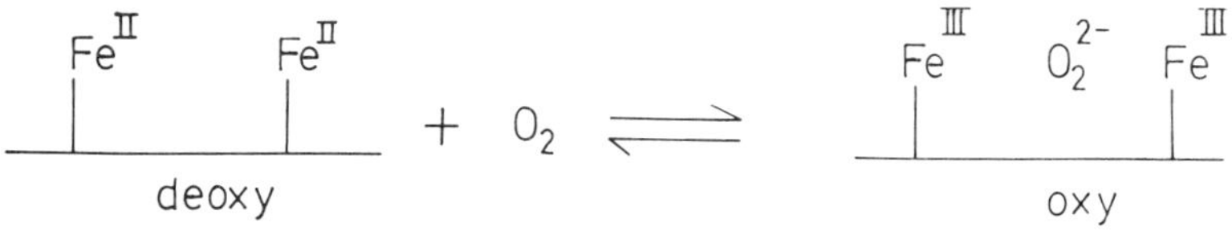

The Mössbauer parameters classify the iron atoms as high spin Fe(II) in deoxyhaemerthyrin and as high spin Fe(III) in oxyhaemerthyrin. The magnetic susceptibility of deoxyhaemerthyrin at room temperature is consistent with four unpaired electrons per iron, as expected for high spin Fe(II).

HAEMOVANADINS

The vacuoles in the blood cells (vanadocytes) of the ascidians (sea squirt) contain a vanadium protein complex dissolved in 0.75-1 M-H_2SO_4. There has been some controversy as to whether this compound can act as an O_2 carrier. Carlisle [17] has claimed reversible uptake of O_2 with a P_{50} value of 2 Torr.

It is rather doubtful if any vanadium-protein complex would be stable under the high acidities reported. Although complexes of various ligands with vanadium(III) and vanadium(IV) or oxovanadium(IV) are known, few complexes with amino acid derivatives have been studied. Recently the preparation of bis(ethyl cysteinato)oxovanadium(IV) (**5.10**) was described; both *cis*- and *trans*-isomers occur.

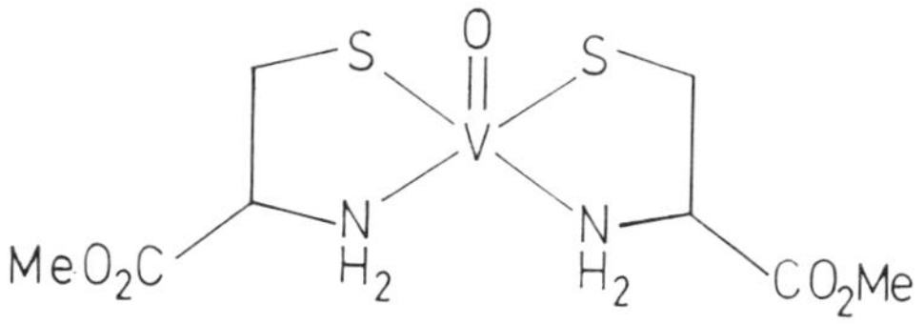

(5.10)

X-Ray absorption spectroscopy has been used to study the vanadium complex in living ascidian blood cells. Analysis of the X-ray absorption edge data shows that only a small amount (<10%) of the vanadium is present as the VO^{2+} ion and the vanadium complex in the living cells is simply aquo-vanadium-(III). Obviously more work is necessary to define this system fully.

SYNTHETIC MODELS FOR OXYGEN BINDING

Although a variety of transition metal complexes are known to bind O_2, interest has largely centred on iron and analogous cobalt complexes. The former being of foremost biological interest and the latter appearing to function as excellent general models for metal-dioxygen binding.

In the case of cobalt(II), the dioxygen adds via an internal redox reaction:

$$\mathrm{LCo^{II}} + \mathrm{O_2} \overset{K_1}{\rightleftharpoons} \mathrm{LCo^{III}O_2^{\cdot -}}$$

$$\mathrm{LCo^{III}O_2^{\cdot -}} + \mathrm{LCo^{II}} \overset{K_2}{\rightleftharpoons} \mathrm{LCo^{III}O_2^{2-}\,Co^{III}L}$$

in which the metal is formally oxidized and O_2 formally reduced. In 1:1 complexes O_2 is formally reduced to the superoxide radical anion $O_2^{\cdot -}$ and in 2:1 complexes to the peroxide ion O_2^{2-}. Table 5.2 lists bond distances for some diatomic oxygen species.

Table 5.2 – Bond lengths of some diatomic oxygen species.

Species	O-O bond length (Å)
O_2^{+}	1.12
O_2	1.20
O_2^{-} (superoxide)	1.32
O_2^{2-} (peroxide)	1.49

Data from A. G. Sykes and J. A. Weil, *Prog. Inorg. Chem.*, **13**, 1 (1970).

The μ-peroxo complexes are generally more stable, and are preferred unless inhibited sterically. The use of non-aqueous solvents (e.g. DMF) and low temperatures may lead to 1:1 complexes.

Dioxygen will react with $[Co(NH_3)_5OH_2]^{2+}$ according to the equation

$$[\mathrm{Co(NH_3)_5OH_2}]^{2+} + \mathrm{O_2} \underset{k_{-1}}{\overset{k_1}{\rightleftharpoons}} [\mathrm{Co(NH_3)_5O_2}]^{2+}$$

with $k_1 = 2.5 \times 10^4\ M^{-1}\ s^{-1}$ at 25 °C. The mononuclear oxygen complex then reacts with the remaining pentaammineaquo complex to give the final binuclear cobalt complex:

$$\mathrm{Co(NH_3)_5O_2]^{2+}} + [\mathrm{Co(NH_3)_5OH_2}]^{2+} \underset{k_{-2}}{\overset{k_2}{\rightleftharpoons}} \underset{\text{(brown)}}{[\mathrm{(NII_3)_5Co.O_2.Co(NH_3)_5}]^{4+}}$$

The overall process, with an equilibrium constant of $6.3 \times 10^5\ M^{-1}$ is reversible over a short period of time (*ca* 10 min), before decomposition to give mononuclear cobalt(III) complexes occurs. The crystal structure of $[(NH_3)_5\ CoO_2\ Co(NH_3)_5]^{4+}$ has been determined, (**5.11**). The O-O bond length of 1.47 Å is fully consistent with that of O_2^{2-}.

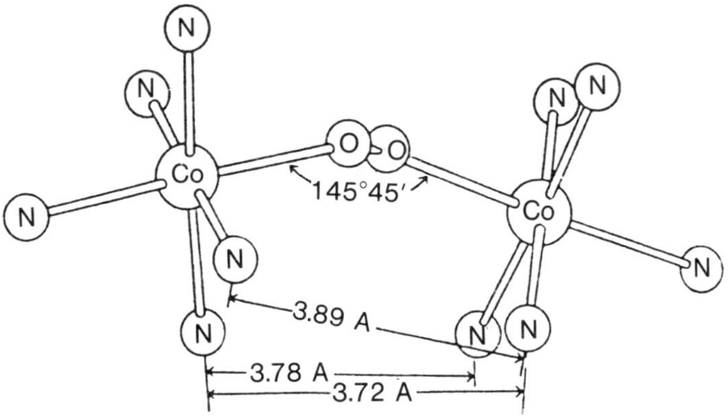

(5.11)

In the case of the μ-peroxo complexes, a second μ-hydroxo bridge is spontaneously formed in aqueous solution whenever a site *cis* to the dioxygen bridge is readily available. The 12-membered tetradentate ligand cyclen **(5.12)** has a relatively small 'hole size' and as a result gives only *cis*-complexes with cobalt(II) and cobalt(III). Reaction of O_2 with cobalt(II)-cyclen gives $[(CoL)_2(O_2)(OH)]^{3+}$ **(5.13)**.

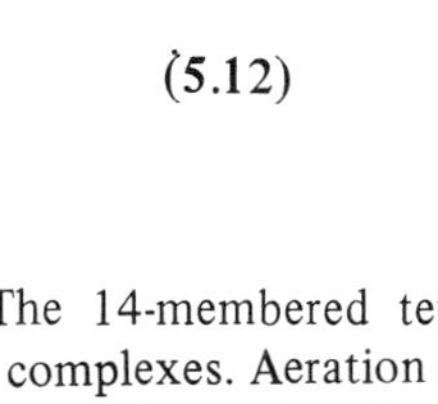

(5.12)

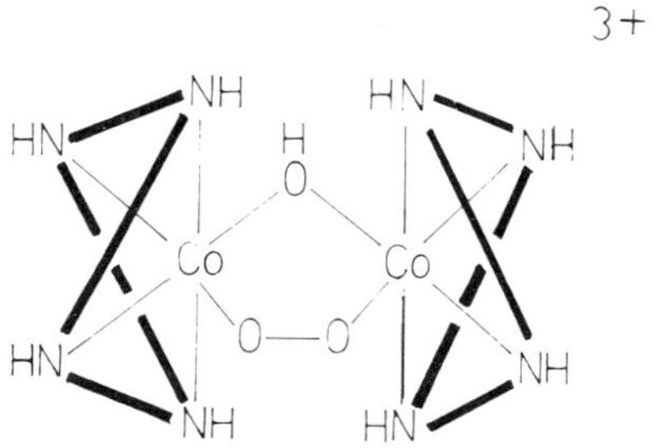

(5.13)

The 14-membered tetradentate ligand cyclam **(5.14)** gives predominantly *trans* complexes. Aeration of the cobalt(II) complex of the ligand in aqueous solution gives $[(CoL)_2(O_2)(H_2O)_2]^{4+}$ **(5.15)**. The axial water ligands may be replaced by various anionic ligands such as Cl^-, Br^-, N_3^- and NO_2^-. A variety of μ-peroxo complexes of this type have been isolated and characterized. Such complexes breakdown in acid solution, but the products have not been fully characterized.

Many problems remain in the area of macrocyclic cobalt(II) complexes. The stereochemistry of the complexes is not well defined in aqueous solution, and it appears to be possible to obtain four, five- and six-coordinate complexes.

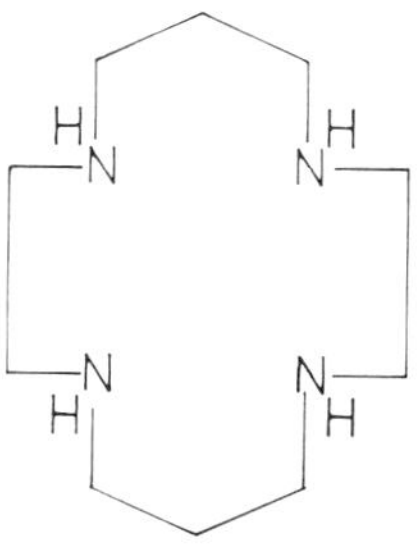

(5.14)

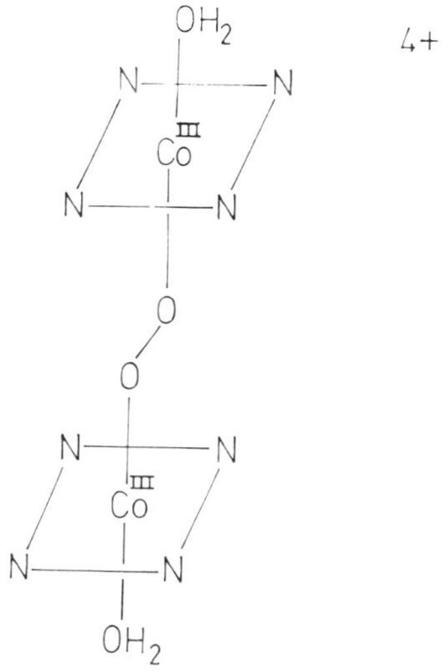

(5.15)

Cobalt(II) complexes of ligands such as Me_6[14] diene (**5.16**) are yellow low-spin species before oxygenation, while cobalt(II) complexes of [15] aneN_4 (**5.17**) and [16] aneN_4 (**5.18**) are red high-spin species. The effect of spin state on oxygenation rates is currently under active investigation.

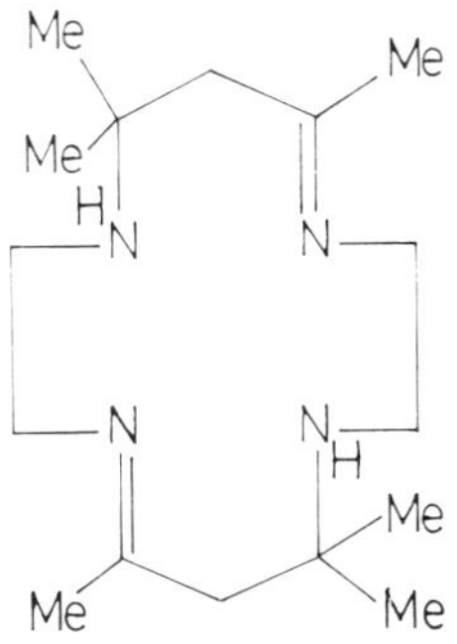

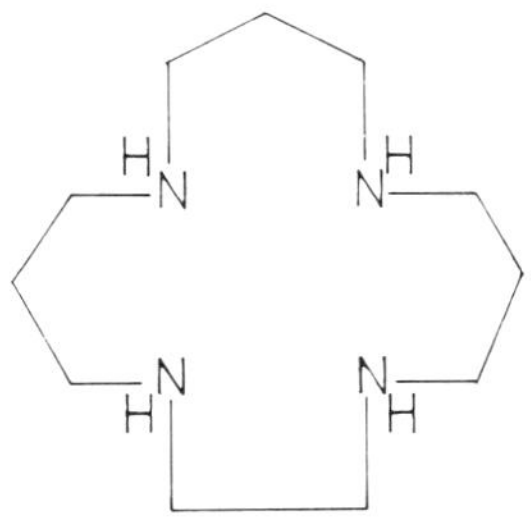

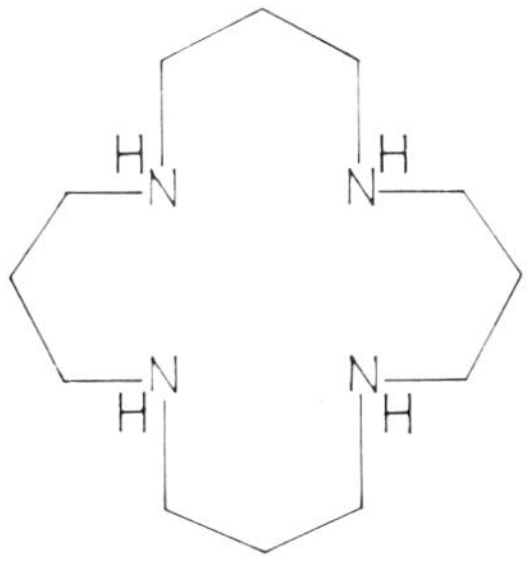

(5.16) (5.17) (5.18)

Very extensive studies have been made of the reaction of O_2 with cobalt(II) complexes of Schiff-base ligands such as acacen (**5.19**) and salen (**5.20**).

The reaction of Co(acacen) in a coordinating solvent such as dimethylformamide, or in a non-coordinating solvent such as toluene with an added base (py or imidazole) at room temperature, results in a non-stoichiometric continuous slow

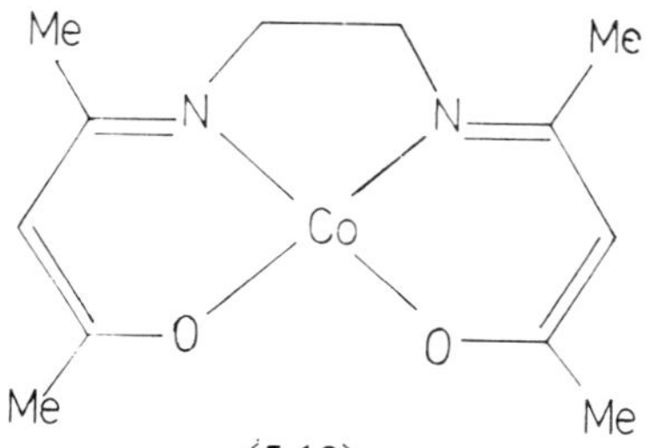

(5.19)

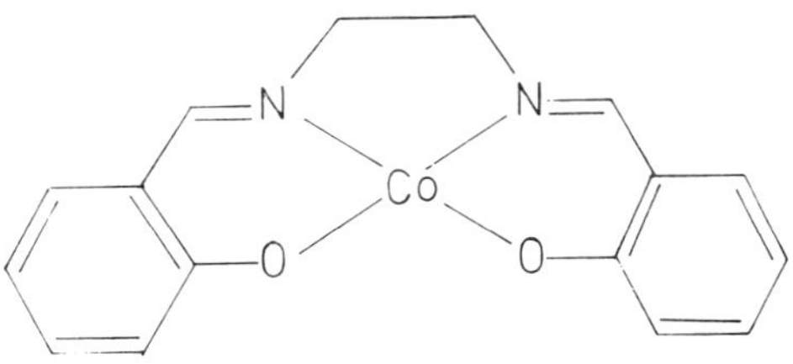

(5.20)

uptake of oxygen over a period of days. Under these conditions it appears that cobalt(II) is catalysing the oxidation of the organic ligand. However, at temperatures near 0 °C or below, there is a rapid and reversible uptake of O_2 corresponding to the formation of the 1:1 $Co(O_2)$ complex:

$$\text{Co(acacen) (B)} + O_2 \rightleftharpoons \text{Co(acacen) (B)} (O_2)$$

where B is the axial base. Various experimental techniques have been used to establish the presence of these 1:1 complexes. The solid complexes have a very strong i.r. band in the region of 1140 cm^{-1} (which disappears when O_2 is removed) attributed to the O–O stretching vibration.

The e.s.r. spectra in both liquid and frozen solutions exhibit eight line hyperfine splitting indicating interaction of the electron with a single ^{59}Co nucleus ($I = 7/2$). There is an ~ 90% transfer of spin density from Co(II) to O_2 indicating that the 1:1 cobalt dioxygen should be formally described as Co^{III}-O_2^-.

An X-ray study of a [Co(Schiff Base) (py)(O_2)] complex has established that the Co–O–O bond angle is 125° (**5.21**). In addition, the O–O distance of 1.2 Å is close to that of superoxide (1.32 Å) and markedly different from that of peroxide (1.49 Å).

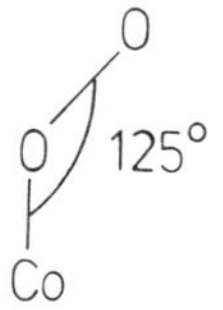

(5.21)

Equilibrium constants K_{O_2} have been determined for the reaction of O_2 with cobalt(II) in many different ligand environments,

$$\text{Co(L) (B)} + O_2 \overset{K_{O_2}}{\rightleftharpoons} \text{Co(L) (B)}\, O_2$$

For a given cobalt(II)-Schiff-base complex Co(L), a linear correlation was found between K_{O_2} and the ease of oxidation of Co(II) to Co(III) as B varies (Fig. 5.5). The basicity of the axial base does not correlate with K_{O_2}, thus l-methylimidazole with a pK_a of 7.25 is much more effective in promoting oxygenation

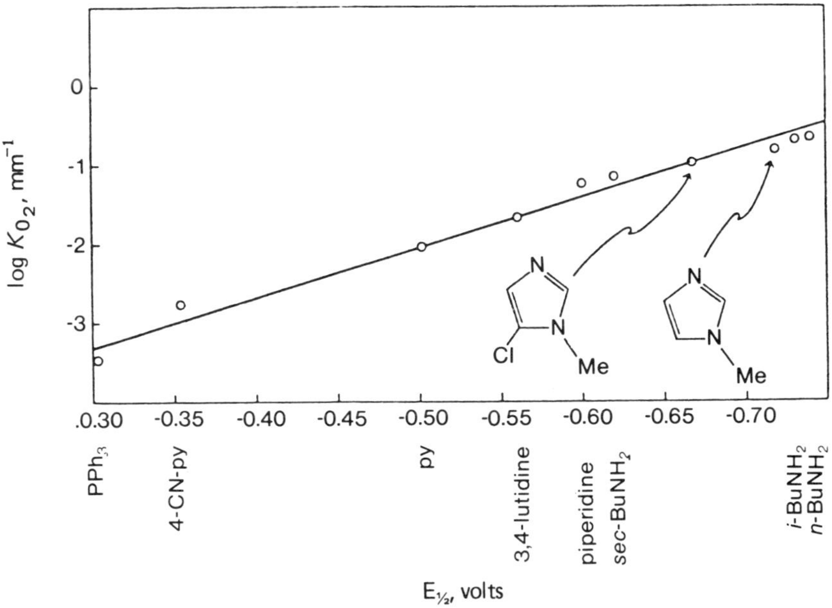

Fig. 5.5 – Reproduced from K. F. Purcell and J. C. Kotz, "Inorganic Chemistry", W. B. Saunders, 1977, by permission of the publishers.

than piperidine (pK_a = 11.3). The effectiveness of the imidazole derivative may arise from its good π-electron donating ability. The bonding of oxygen to a metal complex may be discussed in terms of σ-donation from a non-bonding sp^2 lone pair on oxygen to the d_{z^2} orbital on the metal, accompanied by synergistic π-back bonding from the filled d_{xz} (or d_{yz}) orbitals into the empty π^* orbitals of oxygen. Ligands coordinated *trans* to oxygen will compete for π-electron density on the metal. Good π-acceptors will decrease the π-electron density on the metal resulting in poorer oxygen binding, whereas good π-donors will promote oxygenation by increasing the electron density available for back bonding.

Cobalt(II) haemoglobin and myoglobin reversibly bind oxygen and show many of the features of the iron(II) systems. CoHb retains the full Bohr and phosphate effects on O_2 binding and the binding is cooperative with a Hill constant of at least 2.3. The stereochemical changes that occur as the five-coordinate, low-spin cobalt(II) protoporphyrin IX of CoHb binds oxygen must therefore be compatible with the mechanism which triggers this quaternary structure transition.

Busch and coworkers have recently described some totally synthetic models using the ligand (**5.22**) [10]; The bridge group R′ provides a protected void or 'dry cave' which is intended to shelter small ligands from interactions with other cobalt centres or with solvent. X-Ray data for the cobalt(II) complex isolated as the PF_6^- salt established that cobalt(II) is four-coordinate and square planar. The 'dry cave' is ~ 6.65 Å wide (measured between non-bridgehead vinyl carbons) with the height varying from 4.38 Å (back) to 5.60 Å (front). The saddle shape of the basic macrocycle is largely responsible for the formation of the cavity in the structure. Small ligands can be accomodated in the cave since the cobalt(II) complex can be oxidized in the presence of ligands X^- (NCS^-, N_3^-, NCO^-) to give the six-coordinate $[Co^{III}LX_2]^+$ complexes which are 1:1 electrolytes in CH_3CN.

Reversible u.v.-visible spectral changes are observed for aqueous solutions of the cobalt(II) complex of **(5.22)** ($R' = (CH_2)_6$) containing a large excess of

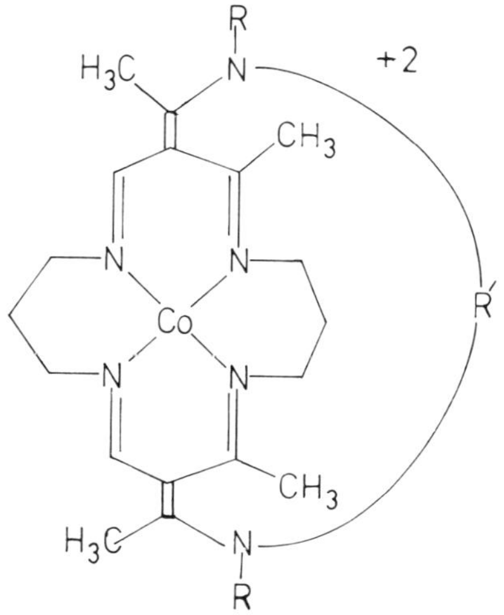

(5.22)

N-methylimidazole, upon oxygenation (Fig. 5.6). The presence of five isosbestic points, coupled with the e.s.r. data, clearly demonstrates a simple equilibrium between the five-coordinate cobalt(II) complex and the 1:1 oxygen adduct.

The P_{50} value for this complex with an axial *N*-methyl imidazole is 0.63 Torr at 20 °C (P_{50} represents the oxygen pressure at half saturation). For comparison at 20 °C, P_{50} for cobalt myoglobin (from sperm whale) is 33 Torr, while P_{50} for iron myglobin (sperm whale) is 0.5 Torr. The results also show that steric effects can easily alter P_{50} by the amounts associated with the change between the T and R states of haemoglobin.

SYNTHETIC MODELS FOR OXYGEN-BINDING HEAMOPROTEINS

In recent years it has been shown that iron(II) porphyrins can react reversibly with dioxygen (equation 5.3) where B is an axial base:

$$Fe(Por)(B)_2 + O_2 \rightleftharpoons Fe(Por)(B)(O_2) + B \quad (5.3)$$

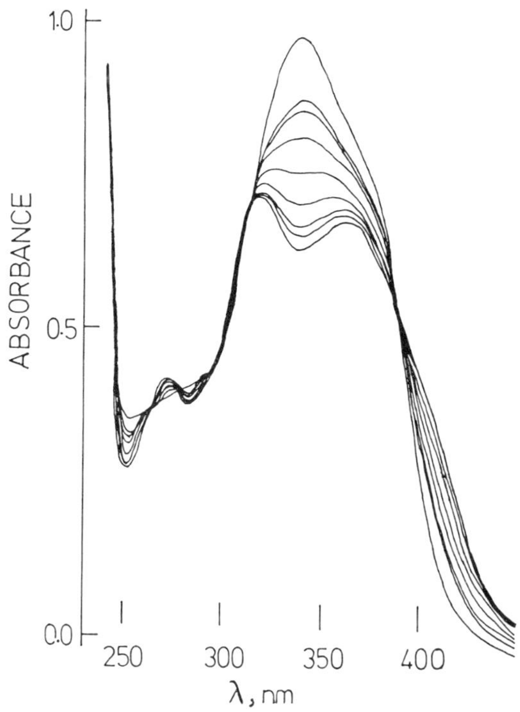

Fig. 5.6 – The electronic spectral changes accompanying formation of the O_2 adduct by the *N*-methyl $(CH_2)_6$-bridged cobalt(II) complex in aqueous solution, containing excess *N*-methylimidazole (2.5 M) at 6 °C. The absorption maximum at 338 nm is due to the O_2 adduct. From reference [10] by permission of the American Chemical Society.

One of the major difficulties encountered in attempts to obtain oxygen carriers based on iron(II) complexes is the large driving force towards the irreversible formation of the μ-oxo dimer (equation 5.4):

$$Fe^{II} + O_2 \rightleftharpoons Fe(O_2) \xrightarrow{Fe^{II}} Fe^{III}\text{-}O\text{-}Fe^{III} \qquad (5.4)$$

The full mechanistic details of this process are not as yet clear, but a schematic representation of the mechanism leading to a μ-oxo dimer in a solid-state reaction is shown in Fig. 5.7.

Considerable research has been aimed at overcoming this problem and three approaches have been successful. These approaches may be summarized; (a) the use of steric contraints in such a way that dimerization is inhibited; (b) the use of low temperatures so that reactions leading to dimerization are very slow, and (c) rigid surfaces – attachment of the iron complex to a surface (e.g. silica gel) so that dimerization is prevented.

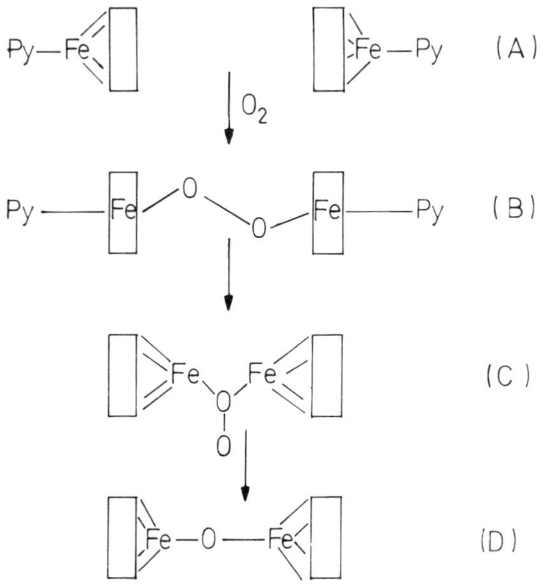

Fig. 5.7 – Schematic representation of the mechanism leading to the μ-oxo complex in a solid-state reaction.

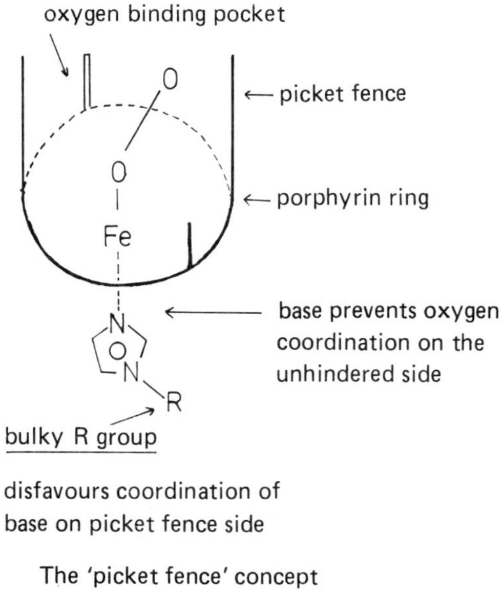

(5.23)

The use of steric contraints has been elegantly demonstrated by Collman [6] and by Baldwin [9]. Similar considerations apply to the Busch macrocycle shown in (**5.22**). In this case formation of a μ-peroxo dimer is inhibited by the 'handle' of the basket type ligand.

Collman developed the concept of the picket fence porphyrins (**5.23**) in order to favour five-coordination and simultaneously inhibit bimolecular reactions. The porphyrin should have great steric bulk on one side of the porphyrin ring, but leave the other side unencumbered. A suitable bulky ligand such as an *N*-alkyl imidazole (which is also an effective π-donor) is coordinated on the unhindered side of the porphyrins thus leaving a hydrophobic pocket for reaction with O_2. The 'picket fence' inhibits bimolecular reactions involving two iron centres and dioxygen.

A suitable picket fence system is shown in (**5.24**). Crystalline diamagnetic dioxygen complexes having 1-methylimidazole on 1-*n*-butylimidazole as axial ligands B have been isolated and characterized. The complexes contain O_2 bound 'end on' (**5.8**) with an Fe–O–O angle of 136° and O–O 1.25 Å. The bonding involved is discussed earlier in the section on 'Iron porphyrins'. A dioxygen complex with tetrahydrofuran as an axial base was also prepared and appears to be paramagnetic (2.4 BM) which could indicate a low spin $Fe^{III}O_2^-$ system.

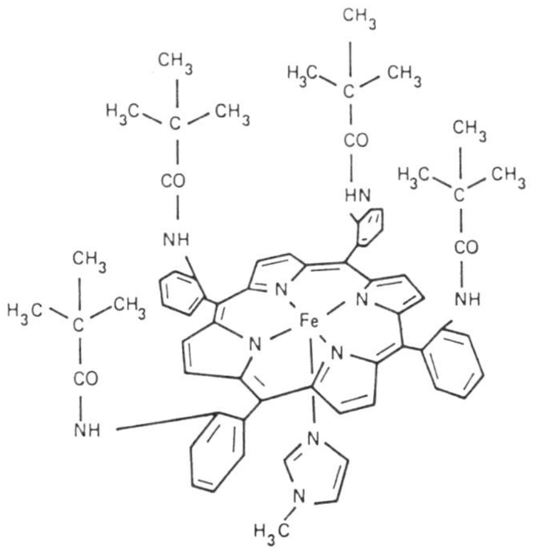

(5.24)

Collman's 'picket-fence' Fe(II)–porphyrin complex [Fe(TpivPP)(1-MeIm)] for reversible O_2-binding; H_2TpivPP is *meso*-tetra (α, α, α, α-*o*-pivalamidophenyl) porphyrin.

The dioxygen complexes show no ν_{O_2} at 25 °C, but a strong very sharp band is seen at 1385 cm^{-1} at −175 °C. Retrospective examination of the 1385 cm^{-1} region in the room-temperature i.r. spectrum reveals a broad weak absorption with a similar integrated intensity to the very sharp low-temperature band. Rapid thermal equilibrium between several rotomeric states which differ slightly in their O_2 stretching frequencies could account for the experimental observations. The 1385 cm^{-1} band is only ~ 100 cm^{-1} below $\nu(O_2)$ ($'\Delta$), a reasonable lowering if there is moderate d_π-p_π back bonding. Formulation of these complexes as low spin Fe(II) and coordinated singlet oxygen is consistent with the i.r. and X-ray data. The various modifications to Fe(II) porphyrins for studies of reversible O_2 binding are shown in Fig. 5.8.

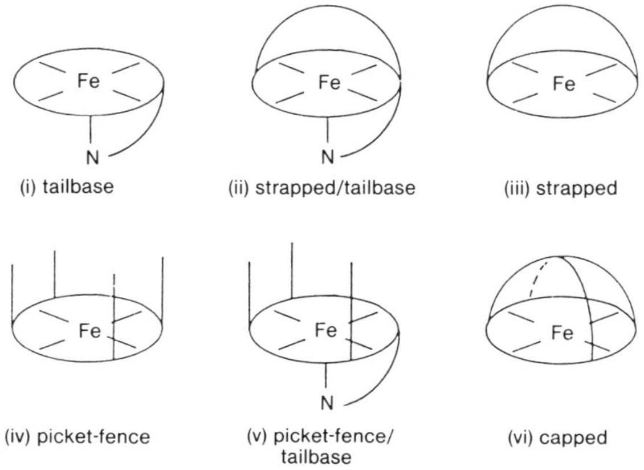

Fig. 5.8 – Different types of modification to Fe(II)-porphyrin complexes so that reversible O_2-binding is possible. From ref [18] reproduced by permission of Academic Press.

Baldwin and coworkers [9] have synthesized the capped porphyrin (**5.25**). The iron(II) complex reacts rapidly with O_2 in pyridine, and deaeration by freeze-thawing restored the spectrum of the iron(II) complex. After several such cycles, little deterioration of the complex was observed. The lifetime of the O_2 adduct in pyridine is *ca* 20 hours. After this time, complete oxidation to the iron(III) complex occurs.

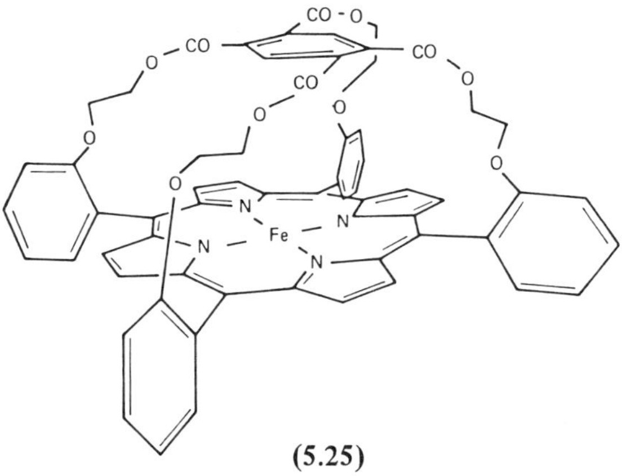

(5.25)

Benzene solutions of the iron(II) complex containing 5% 1-methylimidazole gave evidence for the series of reactions shown in Fig. 5.9.

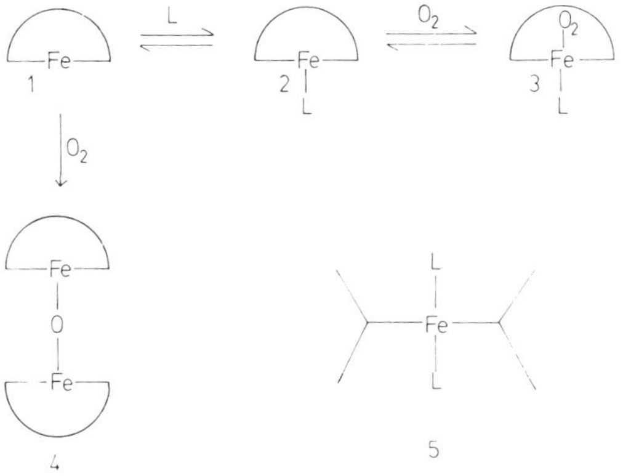

Fig. 5.9 – Reactions of the capped porphyrin.

The stability of the O_2 adduct is largely dependent upon the nature and concentration of the coordinating base L. In the absence of L there is rapid formation of the μ-oxo dimer (Fig. 5.8, (4), while a large excess of L shifts the equilibrium towards the five-coordinate complex with concomitant reduction in the rate of autoxidation. The absence of octahedral species such as (5) which are impossible on steric grounds renders this system less sensitive towards irreversible autoxidation.

One of the first synthetic iron(II) oxygen carriers was an iron porphyrin with a built in imidazole (**5.26**). The fact that it is sterically capable of forming a μ-oxo- dimer and yet reversibly oxygenates at -45 °C suggested that the use

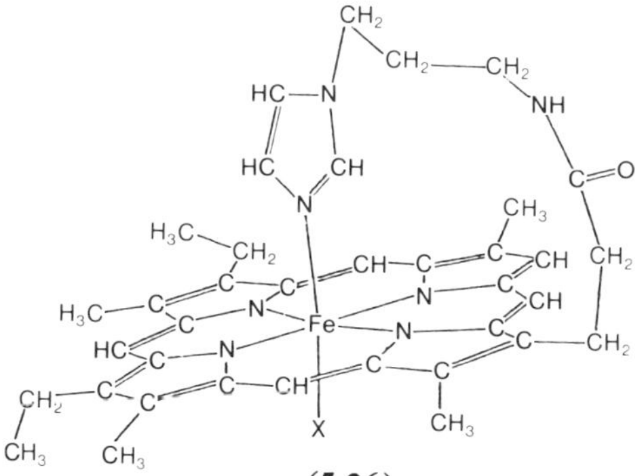

(5.26)

of low temperatures and not the neighbouring group effect of an attached imidazole was responsible. Basolo and coworkers [5] have shown that simple iron(II) porphyrins Fe(TPP)(B)$_2$ (TPP = 5,10,15,20-tetraphenylporphyrin,

(**5.27**) in methylene chloride solution at −79 °C are excellent oxygen carriers. Several oxygenation–deoxygenation cycles are possible at −79 °C without irreversible oxidation occurring. The stoichiometry of the reaction corresponds to equation (5.5).

$$\mathrm{Fe(TPP)\,(B)_2 + O_2 \rightleftharpoons Fe(TPP)\,(B)\,(O_2) + B} \qquad (5.5)$$

Under 1 atmosphere of oxygen, complete complex formation occurs in CH_2Cl_2, but only a small amount of the dioxygen complex is formed in toluene. The behaviour in polar solvents supports formulation of the iron complex as $Fe^{III}(O_2^-)$, and the situation is very similar to that observed with cobalt(II) complexes in polar solvents.

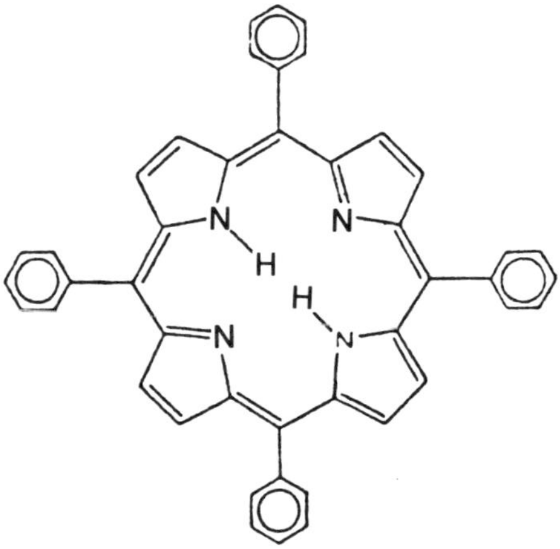

(5.27)

Kinetic studies of the reaction of O_2 and CO with $Fe(TPP)\,(B)_2$ in CH_2Cl_2 at −79 °C are in accord with a rate determining dissociative process (equation 5.6) followed by the rapid reaction of the five-coordinate intermediate, Fe(TPP)B, with either O_2 (equation 5.7) or CO (equation 5.8).

$$\mathrm{Fe(TPP)\,(B)_2} \underset{k_{-1}}{\overset{k_1}{\rightleftharpoons}} \mathrm{Fe(TPP)\,(B) + B} \qquad (5.6)$$

$$\mathrm{Fe(TPP)\,(B) + O_2} \underset{k_{-2}}{\overset{k_2}{\rightleftharpoons}} \mathrm{Fe(TPP)\,(B)\,(O_2)} \qquad (5.7)$$

$$\mathrm{Fe(TPP)\,(B) + CO} \underset{k_{-3}}{\overset{k_3}{\rightleftharpoons}} \mathrm{Fe(TPP)\,(B)\,(CO)} \qquad (5.8)$$

The third successful method for obtaining an oxygen carrying iron(II) complex is to attach it to the surface of a solid so that the two iron atoms cannot approach each other. In a classic experiment Wang [17] reported the first synthetic Fe(II) porphyrin oxygen carrier. Oxygen was found to bind reversibly to 1-(2-phenylethyl)imidazole haem diethyl ester embedded in a matrix of an amorphous mixture of poly(styrene) and 1-(2-phenylethyl)imidazole.

Collman and coworkers [6] prepared Fe(II)(TPP) coordinated to an imidazole group bonded to crosslinked poly(styrene) (**5.28**). Treatment of the polymer-bound complex with O_2 in benzene led to oxidation and formation of the μ-oxo dimer $O[FeTPP]_2$. It was considered that the cross-linked poly(styrene) ligand was not sufficiently rigid to prevent dimerization on treatment with oxygen.

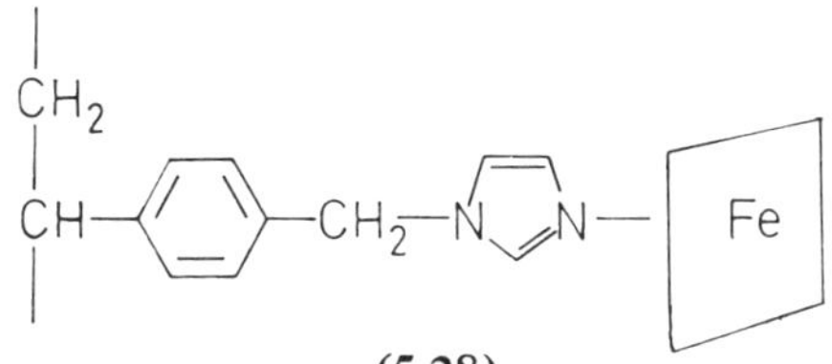

(5.28)

Basolo and coworkers [5] found that attachment of Fe(II) (TPP) to a rigid, modified silica gel support, gave an efficient oxygen carrier. The silica gel used contained a 3-imidazoylpropyl group bonded to the surface of the silicon. Reaction with Fe(II) (TPP) $(B)_2$ followed by heating to remove the axial base gave the five-coordinate complex which reversibly binds O_2 (**5.29**).

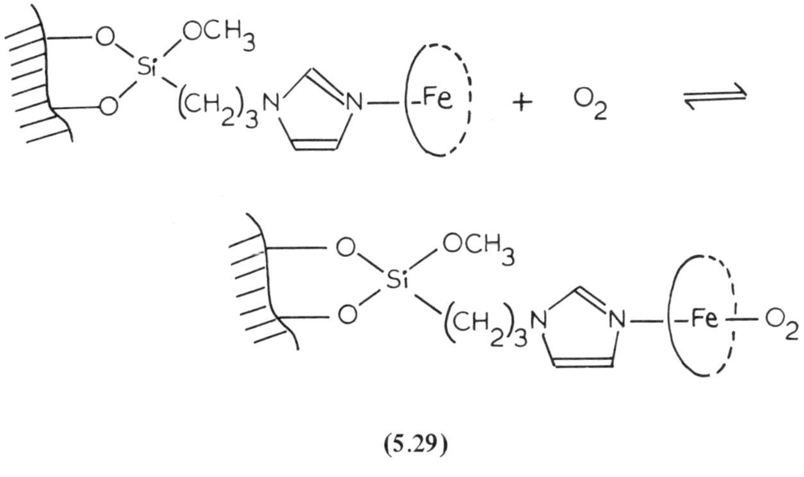

(5.29)

REFERENCES AND BIBLIOGRAPHY

Haemoglobin and Myoglobin

[1] E. Antonini and M. Brunori, *Hemoglobin and Myoglobin in their Reactions with Ligands,* North Holland Publishing Co., Amsterdam, 1971.

[2] M. F. Perutz, *Brit. Med. Bull.,* **32**, 193 (1976). A review of the structure of haemoglobin and the Perutz theory of cooperativity.

[3] R. E. Dickerson and I. Geiss, *The Structure and Action of Proteins*, Harper and Row, New York, 1969.

[4] K. M. Smith, ed., *Porphyrins and Metallophyrins,* Elsevier, Amsterdam, 1975.

Synthetic Models

[5] F. Basolo, B. M. Hoffman and J. A. Ibers, 'Synthetic Oxygen Carriers of Biological Interest', *Acc. Chem. Research,* **8**, 384 (1975).

[6] J. P. Collman, 'Synthetic Models for Oxygen-Binding Hemoproteins', *Acc. Chem. Res.,* **10**, 265 (1977).

[7] G. McLendon and A. E. Martell, *Coord. Chem. Rev.,* **19**, 1 (1976).

[8] J. P. Collman, R. R. Gagne, C. A. Reed, T. R. Hulbert, G. Lang and W. T. Robinson, *J. Am. Chem. Soc.,* **97**, 1427 (1975). J. P. Collman and K. S. Suslick, *Pure Appl. Chem.,* **50**, 951 (1978).

[9] J. Almog, J. E. Baldwin, R. L. Dyer and M. Peters, *J. Am. Chem. Soc.,* **97**, 226 (1975); J. Almog, J. E. Baldwin and J. Huff, *J. Am. Chem. Soc.,* **97**, 227 (1975).

[10] J. C. Stevens, P. J. Jackson, W. P. Schammel, G. C. Christoph and D. H. Busch, *J. Am. Chem. Soc.,* **102**, 3283 (1980).

[11] J. H. Wang, *J. Am. Chem. Soc.,* **80**, 3168 (1958).

Haemocyanins

[12] R. Lontie and R. Witters, in *Inorganic Biochemistry,* ed. G. L. Eichhorn, Vol. 1, Chapter 12, Elsevier, New York, 1973.

[13] R. Lontie and L. Vanquickenborne, in *Metal Ions in Biological Systems,* ed. H. Sigel, Vol. 3, Chapter 6, Marcel Dekker, New York, 1974.

[14] N. M. Senozan, *J. Chem. Ed.,* **53**, 684 (1976).

[15] M. G. Simmons and L. J. Wilson, *J. Chem. Soc. Chem. Commun.,* 634 (1978).

Haemerythrins

[16] D. M. Kurtz, Jr., D. F. Shriver and I. M. Klotz, *Coord. Chem. Rev.*, **24**, 145 (1977).

Haemovanadins

[17] D. B. Carlisle, *Proc. R. Soc. B.*, **171**, 31 (1968).

General

[18] A. G. Sykes, 'Functional Properties of the Biological Oxygen Carriers', *Adv. Inorg. Bioinorg. Mechanisms,* Vol. 2, p. 121. Academic Press, 1982.

CHAPTER 6

Haem Proteins and Copper Proteins in Redox Reactions, Vitamin B_{12}

INTRODUCTION

Redox reactions are fundamental to all life processes. Copper- and iron-containing proteins dominate in the electron transfer role, with the polypeptide or protein component appearing to 'tune' the metal centre to the required redox role. There is also reason to believe that the protein enables electrons to move over considerable distances from one redox centre to another, although the mechanism of this process is still somewhat speculative. Thc redox role of the non-haem iron proteins is discussed in Chapter 7.

BIOLOGICAL REDOX REACTIONS

In general all biological redox systems have redox potentials between that of the hydrogen electrode and that of the oxygen electrode. The hydrogen electrode (NHE) has $E^0 = 0.0$ by definition. This value relates to pH = 0 and must be adjusted to pH 7 as protons are involved in the reaction

$$H^+(aq) + e \longrightarrow \tfrac{1}{2} H_2$$

The potential of the hydrogen electrode at pH 7 ($E^{0\prime}$) can be calculated from eq. (6.1) at 25 °C, and eq. (6.2) at 37 °C.

$$E^{0\prime} = E^0 - 0.059 \times \text{pH} \quad (25\ ^\circ\text{C}) \qquad (6.1)$$

$$E^{0\prime} = E^0 - 0.061 \times \text{pH} \quad (37\ ^\circ\text{C}) \qquad (6.2)$$

At 25 °C, $E^{0\prime} = -0.42$ V.

It is not possible to determine the potential of an oxygen electrode directly as the reaction $\tfrac{1}{2} O_2(g) + 2e \rightarrow O^{2-}(aq)$ is not freely reversible. The e.m.f. of an oxygen–hydrogen cell can be calculated from thermodynamic data.

For the reaction

$$H_2(g) + \tfrac{1}{2} O_2(g) \longrightarrow H_2O\ (l)$$

the value of $\Delta G^0 = -236.60$ kJ mol^{-1} at 25 °C. As $\Delta G^0 = -nFE^0$

$$E^0 = -\frac{\Delta G^0}{nF}$$

$$= \frac{236.6 \times 10^3}{2 \times 96{,}500} = +\ 1.23\ \text{V}$$

(two electrons are involved in the reaction). This value of + 1.23 V relates to pH = 0 and is adjusted to pH 7 using eq. (6.1) giving $E^{0\prime} = +\ 0.82$ V. It is a convention to regard this value as relating to $\frac{1}{2}O_2(g) + H^+(aq) + 2e \rightarrow OH^-(aq)$.

A redox scale can be set up (Fig. 6.1) in which the various $E^{0\prime}$ values are plotted. Hence reduced cytochromes are oxidized by cytochrome oxidase, reduce NADH is oxidized by flavoproteins and reduced cytochrome oxidase is oxidized by O_2.

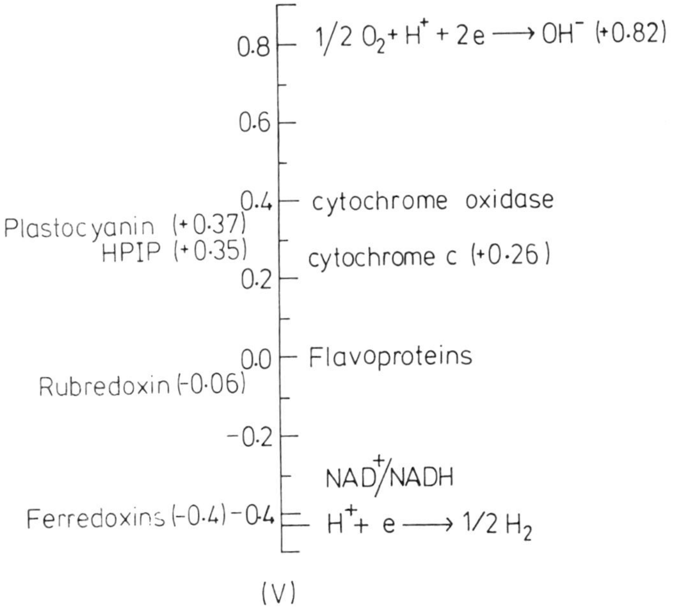

Fig. 6.1 – Redox potentials (V) at pH 7, 25 °C for biological redox systems.

In general each of the redox systems will react only with its more immediate neighbours in the redox potential scale, so that in the oxidation of one by another, the free energy change for the reaction (which is proportional to the distance apart in the scale) is not so great as to render the reaction irreversible.

FACTORS AFFECTING REDOX POTENTIALS IN METAL COMPLEXES

The factors affecting redox potentials of metal complexes in water solvent can be summarized under the headings listed below.

(a) Negative charges in the ligand favour the higher oxidation state; the higher the negative charge, the lower the redox potential. Thus the d^9 copper(II) complex (**6.1**) is readily oxidized to the d^8 copper(III) complex (**6.2**).

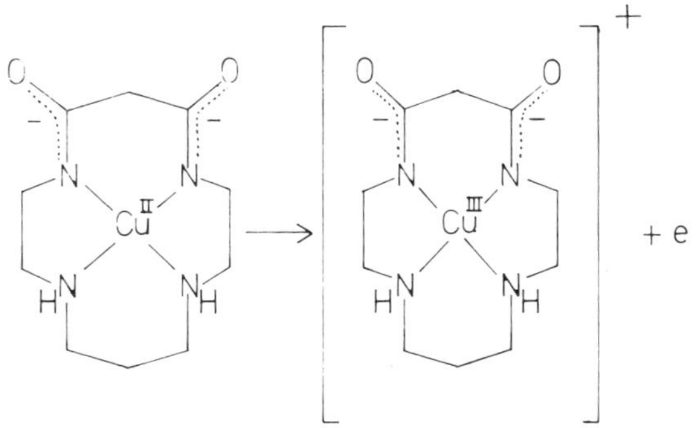

(**6.1**) (**6.2**)

(b) The σ-donor power of the ligand. If the pK_a values are used as a guide to the σ-donor power, a higher pK_a gives a lower redox potential. Strong σ-donors favour high oxidation states.

(c) The greater the π-acceptor power of the ligand, the higher the redox potential (the lower oxidation state is favoured).

(d) Changes of spin state which can alter the importance of (b) and (c) quite markedly. For example. $[Fe(H_2O_6]^{3+}$ has the high spin $t_{2g}^3\ e_g^2$ configuration, while the low spin $t_{2g}^5\ e_g^0$ configuration occurs with $[Fe(CN)_6]^{3-}$. It should be noted that reduction of the negatively charged complex ($E^0 = +\ 0.36V$) is considerably less favoured than that of the aquo ion ($E^0 = +\ 0.71V$). Addition of an electron to a negatively charged species is difficult.

(e) Steric factors, can be chosen so that one or other oxidation or spin state is favoured, so that (b), (c) and (d) can be overridden.

These various points are illustrated by the redox potentials for Cu(II)/Cu(I) couples in Table 6.1.

HAEM PROTEINS

Haem proteins are important in many areas of biology. Haemoglobin and myoglobin act as oxygen carriers, and cytochrome *c* and cytochrome P-450 are involved in redox systems. The chemistry of the globins. (haemoglobin and myoglobin) is discussed in Chapter 5. A structural glossary for the Haems is given in Fig. 6.2.

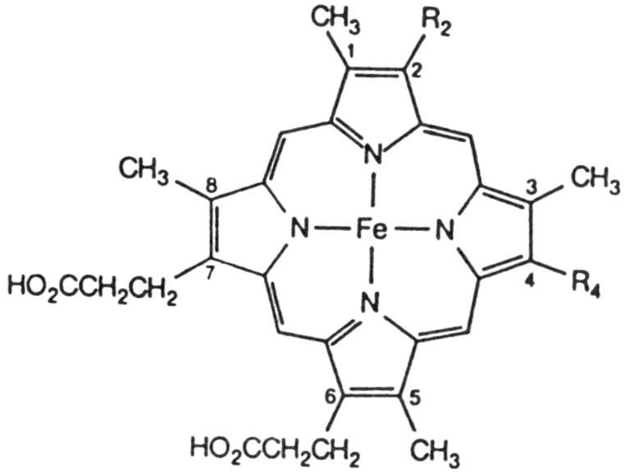

Protohaem	R_2, $R_4 = -CH{=}CH_2$ = Protoporphyrin IX
Synonyms	Fe(II), haem; Fe(III), haematin
Mesohaem	R_2, $R_4 = CH_2CH_3$
Deuterohaem	R_2, $R_4 = H$
Pyrrohaem	R_2, $R_4 = C_2H_5$, H for $CH_2CH_2CO_2H$ at 6
Haem a_2	$R_2 = -CH{=}CH_2$, $R_4 = -CHOHCH_3$
Haem C	R_2, $R_4 = -CHCH_3$ (with $-S-$ attached to CH)
Haem A	$R_2 = CHCH_2CH(CH_2)_3CH(CH_2)_3CH(CH_3)_2$ (with OH, CH_3, CH_3 substituents on the three CH groups)
	$R_4 = -CH{=}CH_2$
	CH_3 at 8 $= -CH{=}O$

haemin: haematin with one axial ligand, for example, Cl^- as in haemin chloride

haemochrome: iron(II) porphyrin with two axial ligands

haemichrome: iron(III) porphyrin with two axial ligands

Fig. 6.2 – A structural glossary for the Haems.

CYTOCHROMES *c*

Cytochromes *c* possess a haem group covalently bound to protein via thioether linkages. Most cytochromes *c* are low spin, the fifth and sixth iron ligands being histidine (His) and methionine (Met) (cytochromes *c* and c_2) or His and His (cytochrome c_3). A few high spin cytochromes *c* are known (cytochrome c') where the axial ligands are probably histidine. The physical properties of some cytochromes are listed in Table 6.2.

Table 6.1 – Redox potentials for some biochemical and chemical couples.

System	E^0 (volts)
Cu (2,9-Me$_2$ 1,10 phen)$_2^{2+}$–Cu (2,9-Me$_2$ 1,10 phen)$_2^+$	+0.59
Cu (2,Cl-1,10 phen)$_2^{2+}$–Cu (2,Cl-1,10 phen)$_2^+$	+0.40
Cu (imidazole)$_2^{2+}$–Cu (imidazole)$_2^+$	+0.35
Cu $(NH_3)_2^{2+}$–Cu $(NH_3)_2^+$	+0.34
Cu (pyridine)$_2^{2+}$–Cu (pyridine)$_2^+$	+0.27
Cu (imidazole)$^{2+}$–Cu (imidazole)$^+$	+0.26
Cu (pyridine)$^{2+}$–Cu (pyridine)$^+$	+0.197
Cu (1,10 phen)$_2^{2+}$–Cu (1,10 phen)$_2^+$	+0.174
Cu^{2+} (aq)–Cu^+ (aq)	+0.167
Cu(ala)$_2^{2+}$–Cu(ala)$_2^+$	−0.130
Cu(gly)$_2^{2+}$–Cu(gly)$_2^+$	−0.160
Laccase Cu^{2+}–Cu^+	+0.415
Ceruloplasmin Cu^{2+}–Cu^+	+0.390
Azurin Cu^{2+}–Cu^+	+0.380
Plastocyanin Cu^{2+}–Cu^+	+0.370

Cytochrome *c* is a remarkably stable enzyme towards both high temperature and extremes of pH. It is reduced readily by dithionite and ascorbic acid. In the reduced form the solution is pink in colour with absorption bands at 551, 522 and 415 nm. The reduced form is only slowly reoxidized by air and *in vivo* the reoxidation depends on the presence of a specific oxidase (cytochrome oxidase). A solution of the oxidized form is yellow with absorption bands at 530 and 400 nm. The redox potential at pH·7 is +0.26 V. The reduction of cytochrome *c* by various inorganic reducing agents has been studied kinetically (Table 6.3).

Cytochrome *c* carries out the deceptively simple role of accepting an electron from cytochrome c_1 and transferring it to cytochrome *c* oxidase. A major objective of cytochrome *c* reasearch (so far largely unattained) is an understanding of its function at the molecular level. X-Ray studies by Dickerson and coworkers [2] have established the essential features of the structures of both oxidized and reduced forms of horse heart cytochrome *c*. Some structural information on ferricytochrome c_{551} from *Pseudomonas aeruginosa* is also available. In horse-heart ferricytochrome *c* [iron(III)], a haem *c* group is bound to the polypeptide chain via covalent sulphur linkages at Cys-17 and Cys-14, as well as by the axial iron ligands His-18 imidazole and Met-80 sulphur (Fig. 6.3).

The haem group is buried in the hydrophobic interior of the protein (Fig. 6.4) except for one edge (shown bold face in Fig. 6.4; rings 2 and 3 in Fig. 6.3) which is near the surface.

Table 6.2 – Physical properties of cytochromes.

Source	Molecular weight	No. of haems	Redox potential (mV)
Cytochrome *c*			
Ox	13500	1	+255
Horse	13500	1	+255
Yeast	13500	1	+260
Turkey	13500	1	+260
Tuna	13500	1	+265
Cytochrome c_1			
Beef	n.d.	1	+220
Cytochrome c_2			
Rhodospirillum rubrum	13000	1	+320
Rhodospirillum molischianum	13400	1	+290
Rhodomicrobium vannieli	n.d.	n.d.	+305
Cytochrome c_3			
Desulphovibrio vulgaris	14700	4	−205
Desulphovibrio vulgaris	26000	8	n.d.
Cytochrome c_4			
Azotobacter vinelandii	24000	2	+300
Cytochrome c_5			
Azotobacter vinelandii	12100	1	+320
Cytochrome c_6 (= f)			
Pea	110000	2	+350
Cytochrome c'			
Rhodopseudomonas palustris	14800	1	+105

Table 6.3 – Rates of electron transfer between cytochromes and small inorganic redox reagents (cy-c = cytochrome *c*).

Reaction	k (M^{-1} s^{-1})	pH	Temperature (°C)
(a) *Protein oxidation*			
Horse cy-*c* + $Fe(CN)_6^{3-}$	1.6×10^7	7.0	25
Beef cy-*c* + $Fe(CN)_6^{3-}$	3.6×10^4	7.4	23
Horse cy-*c* + $Co(phen)_3^{3+}$	1.5×10^5	7.0	25
(b) *Protein reduction*			
Horse cy-*c* + $Fe(CN)_6^{4-}$	2.6×10^4	7.0	
Horse cy-*c* + $Cr(H_2O)_6^{2+}$	1.9×10^4	4.0	25
Horse cy-*c* + $Ru(NH_3)_6^{2+}$	4.2×10^4	7.0	25
Horse cy-*c* + $Fe(edta)^{2-}$	2.8×10^4	3.5-4.5	25

n.d. = not determined.

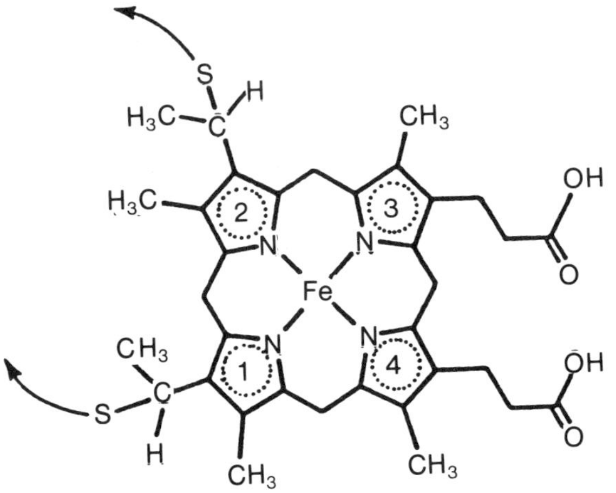

Fig. 6.3 – Structural formula of haem *c*, showing the attachment of the protein from rings 1 and 2.

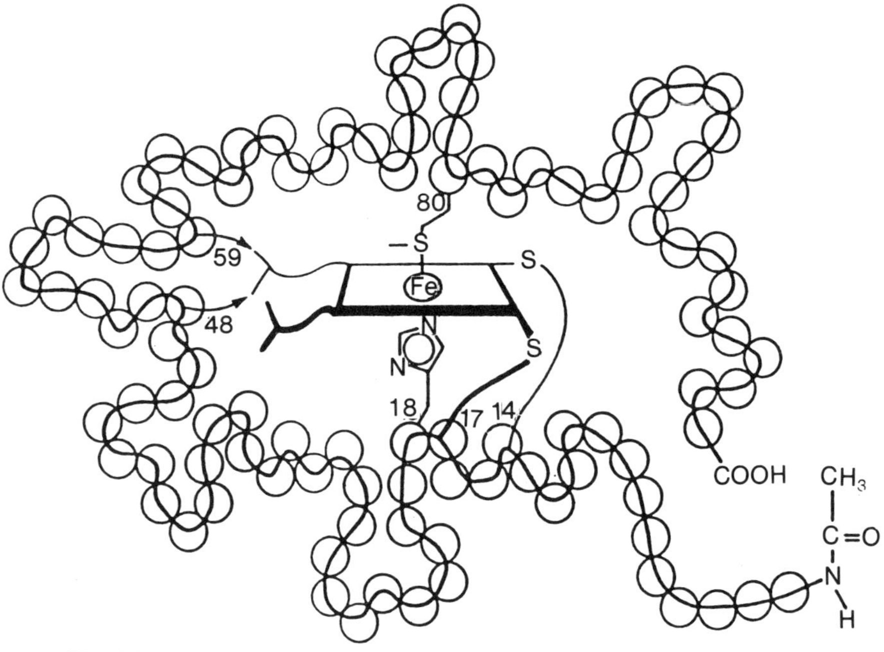

Fig. 6.4A – Schematic representation of cytochrome *c* in the vicinity of the haem.

The 2,3-ring system is referred to as the 'exposed haem edge'. The exact extent of exposure of this edge to solvent molecules and redox reagents is a matter of speculation. Both oxidized and reduced cytochromes *c* possess low spin electronic ground states with configurations t_{2g}^5 and t_{2g}^6. Outer sphere electron transfer of a t_{2g} electron is facile, as minimal reorganizational activation is required.

CYTOCHROME P-450

Cytochrome P-450 is a member of a group of enzymes, which catalyse the addition of molecular oxygen to a substrate via oxygen activation. These haem proteins are monooxygenases, that is that they catalyse the hydroxylation of a substrate RH at the expense of molecular oxygen by the reductive cleavage of the O–O bond.

$$\text{R-H} + O_2 \xrightarrow{2e^-,\ 2H^+} \text{R-OH} + H_2O \qquad (6.3)$$

Dioxygenases insert both atoms of molecular oxygen into the substrate. The 450 designation refers to an intense near-ultraviolet Soret band which is displaced *ca* 30 nm from the corresponding band observed in the spectra of other cytochromes in the carbon monoxide adduct. Cytochrome P-450 is found in plants, animals and bacteria and participates in numerous metabolic pathways. Although the iron-centred prosthetic group and reaction mechanism are believed to be the same in every cytochrome P-450 enzyme, there are many cytochrome P-450's of varying molecular weight (~50,000). Differences in their protein backbones allow this group of monooxygenases to have a wide variety of functions. In humans, different forms of microsomal P-450 are believed to catalyse the hydroxylation of drugs, steroid precursors, pesticides and other foreign substances. Cytochrome P-450 is part of the body's detoxification system, hydroxylating compounds (making them more water soluble) so that urinary excretion is favoured over fat storage. The action of P-450 on certain substrates results in the production of highly reactive intermediates which can then disrupt other cellular components. The carcinogenicity of polycyclic aromatic hydrocarbons has been attributed to their conversion by P-450 *in vivo* to arene oxides.

In general, most P-450 enzymes appear to follow the catalytic cycle shown in Fig. 6.5, where P represents protoporphyrin IX, R the substrate,and R-OH the hydroxylated product. Mercaptide ligation from a cysteine residue has been implicated in the Fe(III) stages (1) and (2) and the Fe(II) + CO (4) stage. It has been suggested that mercaptan (RS^-) may be bound in the Fe(II) deoxy form (3). The idea of a mercaptan/mercaptide proton 'shuttle' operating during catalysis is an intriguing one. Model studies on cytochrome P-450 have been carried out in an attempt to assess the role of the axially ligated sulphur in determining the properties of the active site. These studies may eventually be useful in modelling the oxygen-activating abilities of the enzyme. Collman and

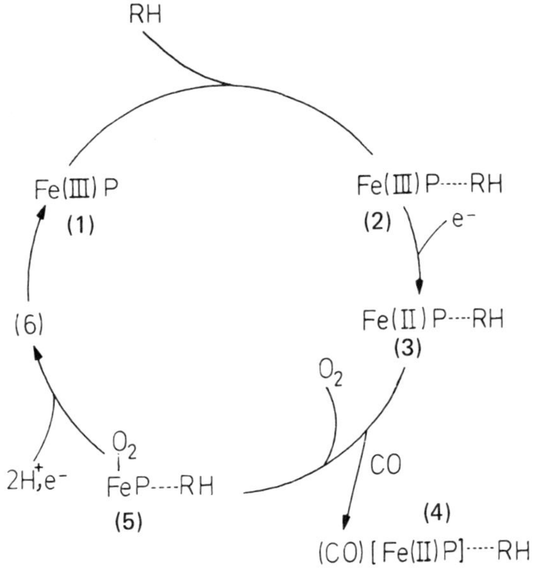

Fig. 6.5

coworkers have synthesized 'mercaptan-tail' porphyrins of type (**6.6**). By the addition of base, six-coordinate mercaptide-Fe(II)-CO complexes can be generated which compare well with the characteristic absorption and MCD spectra of P-450.

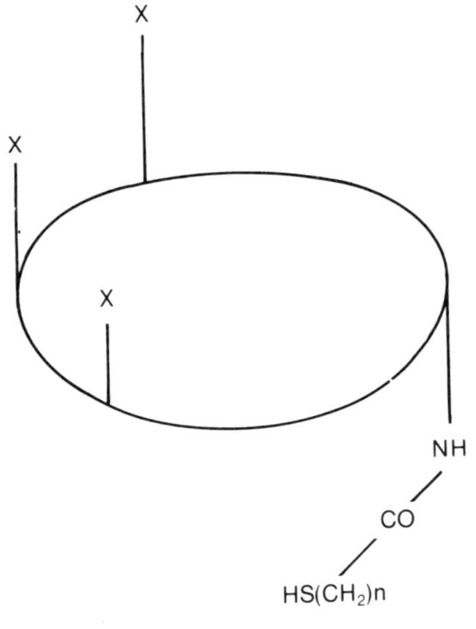

X=NHCOC(CH_3)$_3$; n=5

(6.6)

CATALASE, PEROXIDASE AND SUPEROXIDE DISMUTASE

One of the requirements for life is the maintenance of molecules in a reduced state, even though they are exposed to an oxidizing atmosphere. Another requirement is the generation of energy accomplished by respiration, in which the essential step is the reduction of oxygen to water.

$$O_2 + 4H^+ + 4e \longrightarrow 2H_2O$$

This reaction is catalysed by a single enzyme, cytochrome oxidase. The overall result of this process (Fig. 6.1), is that electrons (accompanied by protons at low potentials) are transferred in stages from the reduced pyridine nucleotides to oxygen, that is through a potential difference of 1.1 V. Theoretically the gain in free energy from the cell is $nFE = 1.1 \times 2 \times 96{,}500 = 212$ kJ. The reduction of O_2 can occur in stages

$$O_2 + e \rightleftharpoons O_2^{\dot{-}}$$
$$O_2^{\dot{-}} + e \rightleftharpoons O_2^{2-}$$

to give superoxide initially, which is a radical anion, followed by the peroxide dianion. These are highly reactive and toxic species, and they can be removed efficiently by the enzymes superoxide dismutase, catalase and peroxidase.

SUPEROXIDE

The chemistry and reactivity of superoxide ($O_2^{\dot{-}}$) has been a subject of considerable interest to biochemists and chemists. In 1965–66 it was first established that dioxygen (O_2) in aprotic solvents is reduced by a reversible one-electron process to superoxide ion

$$O_2 + e^- \rightleftharpoons O_2^{\dot{-}} \qquad E^0 = -0.50 \text{ V vs NHE}$$

which in turn is reduced by a second one-electron process (with involvement of the solvent dimethylsulphoxide)

$$O_2^{\dot{-}} + e \longrightarrow HO_2^- \qquad E_{Me_2SO} = -1.75 \text{ V vs NHE}$$

Stable solutions of $O_2^{\dot{-}}$ in aprotic solvents can be prepared in known concentration by controlled potential coulometry. In the solid state the O–O bond length in $O_2^{\dot{-}}$ is 1.33 Å which corresponds to a bond order of 1.5. The i.r. stretching frequency for $O_2^{\dot{-}}$ is 1145 cm^{-1} compared with 1556 cm^{-1} for O_2 and *ca* 770 cm^{-1} for O_2^{2+}. In acetonitrile with 0.1 M-tetrapropylammonium perchlorate, $O_2^{\dot{-}}$ has a single absorption band with λ_{max} at 255 nm ($\epsilon = 1460$ M^{-1} cm^{-1}). Frozen glasses of the same solution at 77 K give e.s.r. spectra for $O_2^{\dot{-}}$ with $g_\perp = 2.008$ and $g_\parallel = 2.083$. In DMSO, $O_2^{\dot{-}}$ from KO_2 plus [18] crown-6 has an absorption maximum at 250 nm ($\epsilon = 2686$ M^{-1} cm^{-1}). Although $O_2^{\dot{-}}$ is unstable in aqueous solutions, transient amounts can be produced by pulse radiolysis of oxygenated solutions of formate ion. The $O_2^{\dot{-}}$ ion is a strong base:

$$2O_2^{\dot{-}} + HB \longrightarrow O_2 + HO_2^- + B^-$$

and superoxide solutions can promote proton transfer from substrates to an extent equivalent to that of the conjugate base of an acid with $pK_a = 2.3$. The ion is also a nucleophile:

$$O_2^{\dot{-}} + RX \longrightarrow RO_2^{\dot{-}} + X^-$$

BOVINE SUPEROXIDE DISMUTASE

Bovine superoxide dismutase (BSOD), is a dimeric enzyme with a molecular weight of 32,000 containing two Cu(II) and two Zn(II) ions. Its biological role is to catalyse the dismutation of superoxide into O_2 and H_2O_2:

$$2\,O_2^{\dot{-}} + 2H^+ \longrightarrow H_2O_2 + O_2$$

The X-ray structure of the enzyme at 3 Å resolution is now available. The two metals share a common bridging imidazolate residue from the side chain of histidine-61 (Fig. 6.6).

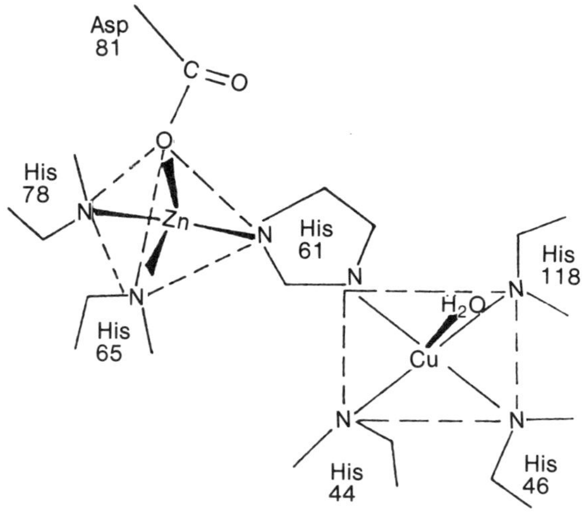

Fig. 6.6 – Schematic drawing of the active site of bovine superoxide dismutase.

The coordination around the copper ion is distorted square pyramidal, the other donors being three histidines and a water molecule. The coordination about zinc is pseudotetrahedral, the four donors being represented in addition to the bridging imidazolate by two histidines and one aspartyl carboxylate residue. The enzyme can be regarded as a heterodinuclear coordination compound.

The two ions can be removed to give a stable apoprotein, which in turn can reaccept different ions. In such a way it has been possible to prepare a large variety of metallo-substituted superoxido-dismutases (Table 6.4). At the present time only Ag(I) and Co(II) have been inserted in place of the native copper(II), whereas a larger number of metals have been substituted in place of the native zinc ion. The zinc-depleted enzyme is labelled Cu_2E_2BSOD.

Table 6.4 – Relative activity of metal-substituted bovine superoxide dismutases.

Native enzyme	100
Apoprotein	0
Cu_2Co_2BSOD	90
Cu_2Co_2BSOD	90
Cu_2Cd_2BSOD	70
Cu_2Cu_2BSOD	100
E_2Co_2BSOD	0
E_2Zn_2BSOD	0
$Ag(I)_2Cu_2SOD$	5

The copper ion in the native enzyme represents the catalytic centre, the zinc ion has only an ancillary structural function. Every derivative which maintains copper(II) in the native site retains almost full activity, whereas no derivative in which the native copper is replaced by other metals retains this activity. The copper atom also has a structural role and stabilizes the enzyme quite strongly.

Addition of $O_2^{\dot{-}}$ to the fully oxidized enzyme causes a bleaching of the colour, whereas addition of superoxide to the fully reduced enzyme partially restores the colour. This behaviour indicates that the copper ion is alternatively reduced and oxidized during the catalytic cycle. A possible mechanism is the following:

$$Cu(II)\,(His^-)\,(HisH)_3 + O_2^{\dot{-}} \xrightarrow{H^+} Cu(I)\,(HisH)_4 + O_2$$

$$Cu(I)\,(HisH)_4 + O_2^{\dot{-}} \xrightarrow{H^+} Cu(II)\,(His^-)\,(HisH)_3 + H_2O_2$$

Recent ^{113}Cd-n.m.r. investigations on Cu_2Cd_2SOD suggest that the histidine becomes protonated on the copper site during the catalytic cycle.

PEROXIDASES

Peroxidases are enzymes catalysing the oxidation of a variety of organic and inorganic compounds by H_2O_2 or related compounds. All the peroxidases purified from plants contain the haemin group (Fe(III)-protoporyphyrin IX). Horseradish roots and the sap of fig trees are the richest source of plant peroxidases. The properties of some peroxidases are listed in the Table 6.5. Most peroxidases are glycoproteins, but the purpose of the carbohydrate in these enzymes is uncertain. A typical peroxidase catalysed reaction is:

$$H_2O_2 + AH_2 \xrightarrow{\text{peroxidase}} 2\,H_2O + A$$

Table 6.5 – Some properties of the peroxidases.

	Molecular weight	Prosthetic group	Carbohydrate content (%)	Crystallization
Horseradish peroxidase (HRP)	40,500[a]	Ferriprotoporphyrin IX	18.4	microscopic needles
	39,800[b]		18	
Cytochrome *c* peroxidase (C*c*P)	34,100[b]	Ferriprotoporphyrin IX	0	long prisms
Chloroperoxidase (CIP)	40,200[a]	Ferriprotoporphyrin IX	25–30	brown needles
	42,000[b]			
Lactoperoxidase (LP)	76,500[a]	Derivative of mesohaem IX?	8	needles
	77,500[b]			
Thyroid peroxidase (ThP)	62,000[c]	Not ferriprotoporphyrin IX?	–	–
Japanese radish peroxidase *a* (JRP*a*)	55,700[f]	Ferriprotoporphyrin IX	28	rhombic prisms
	55,500[d]			
Japanese radish peroxidase *c* (JRP*c*)	41,500[a]	Ferriprotoporphyrin IX		tetragonal prisms
Myeloperoxidase (MP)	149,000[b]	Two atoms of porphyrin bound iron	0	needles
NADH peroxidase	12,000[e]	FAD	–	–
Turnip peroxidase (TuP) A_1	49,000[d]	Ferriprotoporphyrin IX	–	fine needles
A_2	45,000[d]			
B	65,000?[d]			
D	43,000[d]			
Glutathione peroxidase	90,000[d]	One atom of Se per subunit	–	–

[a]Haem content. [b]Hydrodynamic measurement. [c]Gel filtration. [d]Chemical analysis. [e]FAD content. [f]Osmotic pressure.

Horseradish peroxidase (HRP) has long been thought to involve a histidine residue and an aquo group as the fifth and sixth ligands to Fe(III). In solutions of low pH, HRP is high spin, but a low spin species is formed at high pH (*ca* 11).

Two equivalent oxidation of Fe(III) HRP with H_2O_2 gives a green complex (HRP-I), which contains Fe(IV) and a radical species (the extra electron is removed from the protein or porphyrin). The addition of one electron to HRP-I gives the non-radical iron(IV), HRP-II. Recent n.m.r. investigations suggest that HRP-I and HRP-II are high- and low-spin iron(IV) species and that the radical species is not associated with the haem group. Horseradish peroxidase may be reduced by dithionite to give an iron(II) peroxidase which reacts with O_2 to give an inactive species (HRP-III).

The reactions involved with peroxidase are of the type:

$$\text{HRP} + H_2O_2 \longrightarrow \text{HRP-I}$$

$$\text{HRP} + AH_2 \longrightarrow \text{HRP-II} + AH\cdot$$

$$\text{HRP-II} + AH_2 \longrightarrow \text{HRP} + AH\cdot$$

leading to the formation of free radicals which then react further, for example with O_2 to give superoxide ion.

CATALASES

Catalases and peroxidases are related enzymes in so far as they are both capable of promoting hydrogen peroxide oxidation by mechanisms which involve similar enzymic intermediates. The *peroxidatic* activity of catalases is low when compared with true peroxidases. The *catalatic* activity of catalases is the highly efficient catalysis of the decomposition of H_2O_2 to water and oxygen:

$$2\,H_2O_2 \longrightarrow 2\,H_2O + O_2$$

The catalases have molecular weights of about 240,000, made up of four identical subunits, each containing one haem group, with high spin iron(III). Chemical modification studies have suggested that histidine and tyrosine residues are involved in the activity of the enzyme. The axial metal sites appear to be occupied by water and an amino-acid residue.

Catalase reacts with H_2O_2 to give compound I, which is then able to oxidize hydrogen peroxide. As in the case of peroxidase, compound I is believed to be iron(IV) with a radical group in the protein or porphyrin.

The reaction of Fe(III) with H_2O_2 has been used as a model for catalase activity. By using ^{18}O labelling, it has been shown that both oxygens in the evolved O_2 molecule originate from the same H_2O_2 molecule. Kinetic and spectrophotometric studies led to the mechanism:

$$Fe^{3+} + HO_2^- \underset{k_2}{\overset{k_1}{\rightleftharpoons}} Fe^{3+}\,HO_2^-$$

$$Fe^{3+}\,HO_2^- \xrightarrow{k_3} FeO^{3+} + OH^-$$

$$FeO^{3+} + H_2O_2 \xrightarrow{k_4} Fe^{3+} + O_2 + H_2O$$

One important feature of catalase is that it can catalyse the decomposition of H_2O_2 molecules, while in the model systems catalysis involves HO_2^- (pK_a of H_2O_2 is 11.62). A comparison of the catalytic activity in the series of iron(III)-centred catalysts, $Fe(H_2O)_6^{3+}$, ferrihaem monomer and catalase suggests that the unique feature of the catalase action resides in the pH-independence of the reaction. Indeed at pH 12, extrapolated data suggests that the three catalysts have similar activities. The key feature of catalase lies not in the rate acceleration, but in its ability to use H_2O_2 molecules.

It is interesting to speculate that a histidine residue may act as general base catalyst as in (**6.4**) to form the required HO_2^- ion in catalase.

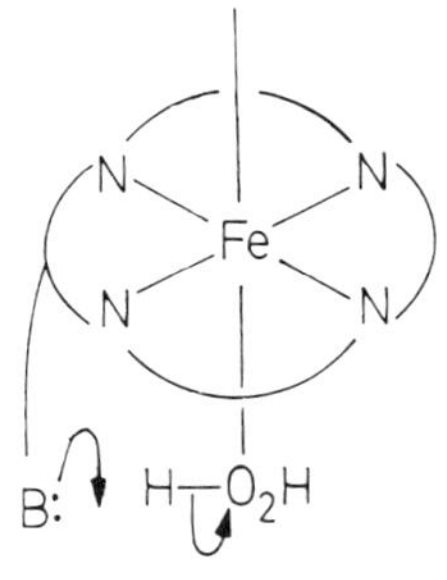

(6.4)

BLUE-COPPER PROTEINS

Copper-containing proteins are involved in a variety of biological functions. These functions include electron transport, copper storage and many oxidase activities. Several copper proteins are easily identified by their beautiful blue colour and have been labelled 'blue-copper' proteins. As shown in Table 6.6, the blue copper proteins can be divided into two classes, the oxidases and the electron carriers.

Copper(II) sites in proteins can be classified into three types based on their spectral properties. The blue (Type I) copper proteins are characterized by a visible absorption band near 600 nm with a high extinction coefficient (Table 6.7).

Table 6.6 – Two classes of blue-copper proteins.

Protein	Molecular weight	No. of Cu	Types of Cu	Source	E'_0 (mV)
Oxidases					
Laccase	60,000–141,000	4	I, II, III	*Rhus vernicifera* *Polyporous versicolour*	450
Ascorbate oxidase	140,000	8	I, II, III	Plants and bacteria	
Ceruloplasmin	132,000	6	I, II, III	Human serum	390
Electron transport proteins					
Plastocyanin	10,500	1	I	Higher plants and cyanobacteria	
Stellacyanin	20,000 (107 a.a)†	1	I	*Rhus vernicifera*	415
Rusticyanin	16,500	1	I	*Thiobacillus ferro-oxidans*	680
Umecyanin	14,600	1	I	Horseradish	
Plantacyanin	8000	1	I	Cucumber and spinach	

†a.a. = amino acid residues.

Table 6.7 – Absorption spectra of copper oxidases.

Copper oxidase	Absorption Max (nm)	Extinction coefficient ($M^{-1}cm^{-1}$)
Laccase [Cu(II)]	730	~500
	615	1400
	532	~300
Azurin (psuedomonas blue Protein) [Cu(II)]	806	~600
	621	2800–3500
	521	~300
	467	~400
Ascorbic acid oxidase [Cu(II)]	606	770
	412	~500
Ceruloplasmin [Cu(II)]	605	1200
	370	~500

As a result, Type I sites are often referred to as blue-copper proteins as their solutions are blue at typical enzyme concentrations employed in the laboratory (10^{-4}-10^{-5} M). The 600-nm band is considered to arise from a L → M charge transfer transition involving a copper–cysteine bond.

Type(II) copper(II) proteins, or low blue copper have less intense colours at normal concentrations, but even low blue-copper sites have quite high extinction coefficients when compared with simple copper(II) coordination compounds. Bovine erthrocyte superoxide dismutase(BSOD) is an example of a low blue copper protein with λ_{max} 680 nm ($\epsilon = 300\ M^{-1}\ cm^{-1}$) Table 6.8.

Table 6.8 – Properties of bovine erthrocyte superoxide dismutase.

Molecular weight	31,400	two identical subunits
Amino acids/subunit	151	isoelectric point (pI) – 4.95
Metals/subunit	1 Cu, 1 Zn	colour – blue-green
Absorption spectrum	λ_{max} (nm)	ϵ_{max} (M^{-1} cm^{-1})
	258	10,300
	270, 282, 289	shoulders
	680	300
E.p.r. spectrum (77 K)	g_m 2.080	$g_{\parallel}$ 2.265
		$A_{\parallel}$ 0.015 cm^{-1}
Redox potential (pH dependent)	$E^{0\prime} = 0.42$ V	

Type III copper(II) found for example in *Rhus laccase* is e.s.r. inactive, i.e. although copper(II) is present no e.s.r. spectrum can be obtained. Recent magnetic susceptibility measurements on *Rhus laccase* indicate an antiferromagnetically coupled copper(II) dimer. Galactose oxidase may be the only non-blue protein which contains a single copper(II). The enzyme which catalyses the oxidation of galactose by molecular oxygen has M = 68,000 and the single copper(II) can be removed using diethyl dithiocarbonate ($HSSCNEt_2$) or H_2S. The apoenzyme is quite stable and the enzyme can be totally regenerated by the addition of Cu(II). Of the six electron transport proteins listed in Table 6.6, the azurins and plastocyanins have received particular attention in structural studies. Plastocyanin is found in the chloroplasts of higher plants, algae and in many cyanobacteria. The function of the protein is to pass an electron from membrane bound cytochrome *f* to the P_{700} chlorophyll reaction centre in the photosynthetic electron transport chain. Azurin appears to be a component of the respiratory electron transport chain in several bacteria. Both proteins are small, Table 6.6 and contain a single type I copper atom. The intense blue colour of the purified proteins has stimulated many structural studies of the copper binding site.

PLASTOCYANIN

The structure of oxidized poplar plastocyanin is shown in Fig. 6.7. The copper ligands are the imidazole nitrogens of His-37 and His-87, and the sulphur atoms of Cys-84 and Met-92. A distorted tetrahedral geometry occurs, the bond angles differing by up to 50° from tetrahedral values. The ligand donors (hard nitrogen and soft sulphur) represent a compromise between the requirements of Cu(I) and Cu(II). The stereochemistry is intermediate between the planar and tetrahedral geometries favoured by Cu(II) and Cu(I), so that rapid electron transfer is expected.

AZURIN

X-Ray and EXAFS data on oxidized azurin show the presence of a very short Cu–S distance of 2.10 ± 0.02 Å, while two nitrogen ligands (imidazole) are 1.97 Å

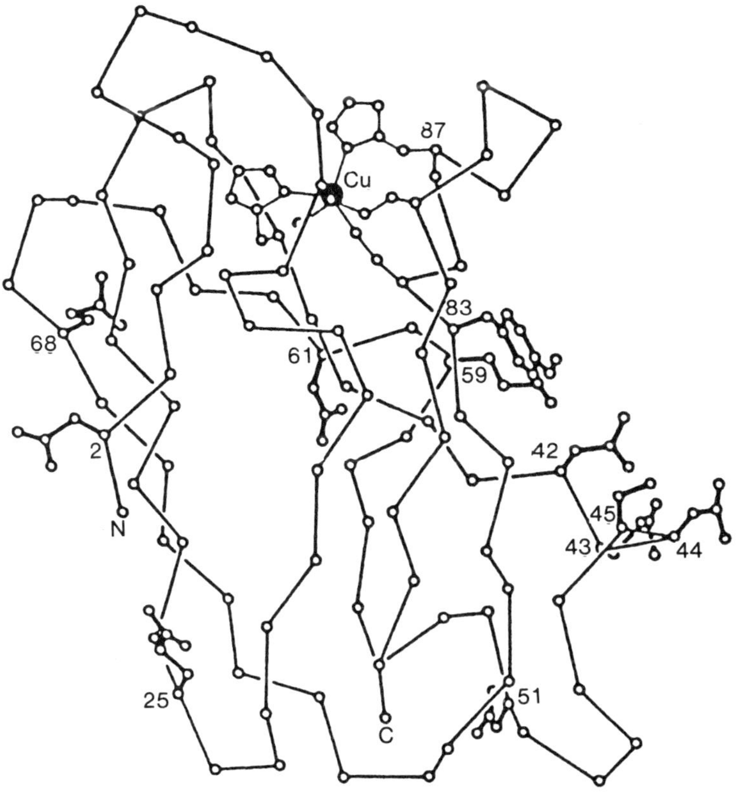

Fig. 6.7 The structure of poplar plastocyanin as reported by Freeman *et al.* [14].

from the metal. The fourth ligand appears to be a sulphur donor Cu–S ~ 2.25 Å (probably methionine). There are therefore close similarities between plastocyanin and azurin.

It is noteworthy that stellacyanin does not contain a methionine residue so that different ligands must occur at this Type I copper site. This variation may be relevant to changes in the redox potential.

BLUE OXIDASES

The most studied examples of the blue oxidases are laccase, ceruloplasmin and ascorbate oxidase. Laccase is a water soluble enzyme and easily purified. The enzyme contains one mole of Type I, one mole of Type II copper, and two moles of antiferromagnetically coupled Type III copper. Laccase, discovered by Yoshida (1883) in the latex of the Japanese lac trees (*Rhus* spp.) is a polyphenol oxidase. The enzyme catalyses the oxidation of *o*- and *p*-dihydroxy phenols to quinones. The pair of e.s.r. non-detectable copper ions are commonly regarded as functioning as a cooperative two-electron unit, resulting in a two electron reduction of O_2 which bypasses the formation of superoxide.

Several complexes have been synthesized as possible models for type III Cu. These complexes generally involve macrocyclic ligands which can bind two copper ions.. The macrocycle (**6.5**) = L gives the complex $Cu_2L(N_3)_4$ with azide and copper(II), with a Cu-Cu separation of 5.145 Å and is completely diamagnetic between 4.2 and 390 K. Although many binuclear copper(II) complexes exhibit antiferromagnetic behaviour, few of them are diamagnetic at room temperature and have smaller Cu-Cu separations. The Curtis macrocycle Me_6[22]aneN_4 (**6.6**) gives the dinuclear complex (**6.7**) with bridging methoxy groups. The Cu-Cu separation is 3.03 Å, and the Cu-O-Cu angle is 102.5°. The magnetic moment is 0.87 BM at 299 K and 0.19 BM at 81 K.

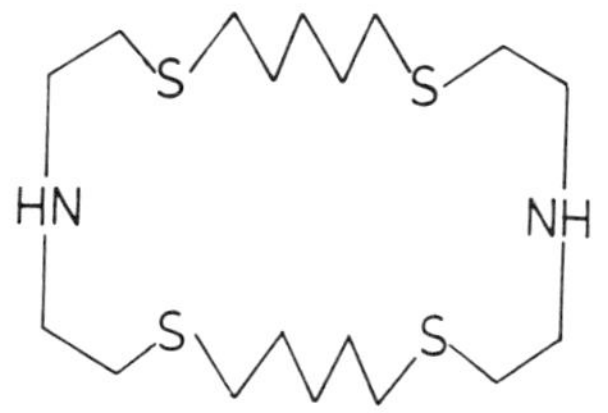

(6.5) = L

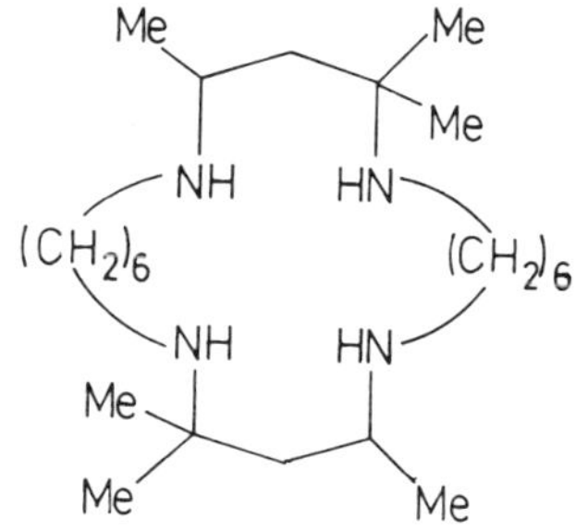

(6.6)

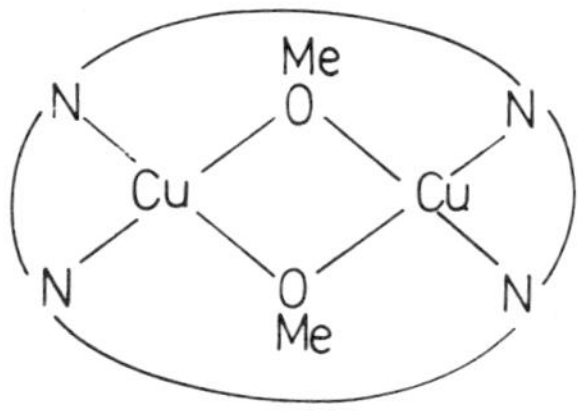

(6.7)

CERULOPLASMIN

Ceruloplasmin has been called the enigmatic copper protein. It is the intensely blue coloured copper containing glycoprotein of the α_2-globulin fraction of mammalian blood. Despite intense research activity over the past 30 years, the physiological role(s) of the protein are not, as yet, known with certainty. The structure, copper content, and nature of the copper-binding sites has also not been fully established. Ceruloplasmin probably has seven or eight copper atoms per mole ($M \sim 132{,}000$) with two Type I, one type II and four Type III sites. Irradiation at 77 K of oxidized ceruloplasmin at the 370 nm band (the Type III chromophore) leads to the reduction of Type I copper (loss of the intense absorption at 605 nm), Type II copper is unaffected. These results indicate that direct energy transfer between Type I and Type II sites is possible in agreement with the role of these sites for substrate and dioxygen binding respectively. Similar results have been obtained with laccase.

SYNTHETIC COPPER COMPLEXES AS ANALOGUES OF THE ACTIVE SITE OF BLUE COPPER PROTEINS

Prior to the X-ray crystallographic work on the copper site of plastocyanin, copper-N (His) and/or copper-S bonds were inferred from a number of studies. Thus Ibers and his collaborators prepared $[Cu^{I}N_3(SR)]$ and $[Cu^{II}N_3(SR)]$ complexes by the reaction of [Cu(SR)] or [Cu(SR)] (ClO_4) (SR = *p*-nitrobenzenethiolate or *O*-ethylcysteinate) with hydrotris-(3,5-dimethyl-1-pyrazolyl)borate (**6.8**) to give complexes of the type shown in (**6.9**).

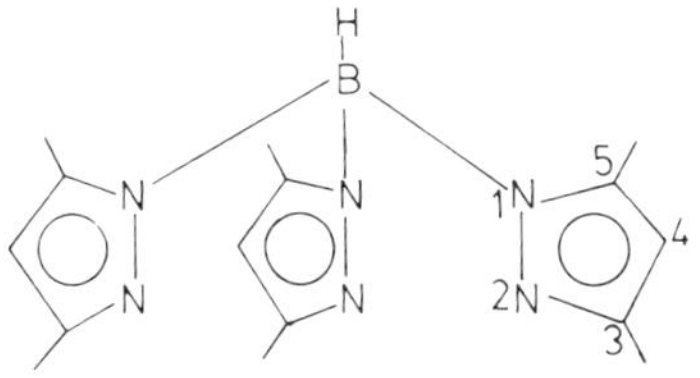

(6.8)

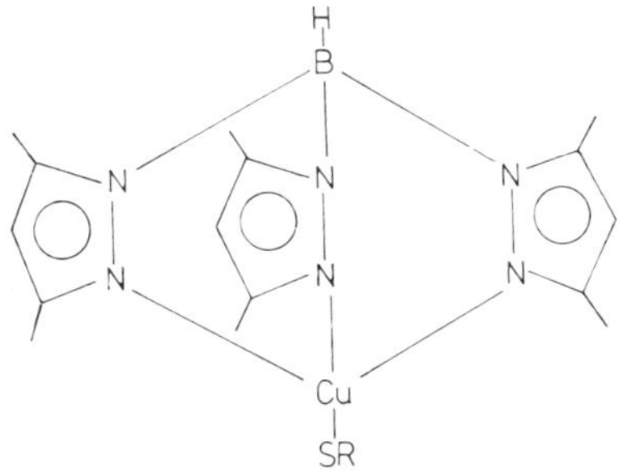

(6.9)

Structure (**6.8**) shows a view of the coordination geometry about the copper atom for such a complex. As Fig. 6.9 shows, the visible spectrum of the $Cu^{II}N_3(SR)$ species is not dissimilar from that of the native system (*Pseudomonas* azurin). However, in marked contrast to the narrow hyperfine couplings seen in th e.s.r. spectrum of the metalloprotein the couplings for $[Cu^{II}N_3(SR)]$ are not unusually small, with $A_{\parallel} = 17.1$ mK when SR = $O_2NC_6H_4S^-$. Sakaguchi and Addison have studied the e.s.r. of $[Cu^{II}S_4]$ chromophores and have suggested that e.s.r. parameters of Type I copper centres are compatible with CuS_2N_2 or $CuSN_3$ chromophores. A series of complexes of the Schiff base (**6.10**) and Cu(II) prepared from pyrrole-2-carboxaldehyde and the appropriate amines have been studied in detail (visible spectrum, e.s.r. and redox potentials).

There are smooth correlations between *d–d* band energies, $A_\parallel$, $g_\parallel$, A_0 and g_0 parameters. As the dihedral angle between the chelate rings increase from 0° to 90°, g and $A_\parallel$ decrease in an antiparallel fashion, while the redox potentials shift to more positive values. These observations are consistent with tetrahedral distortion at the metal binding sites of blue-copper proteins.

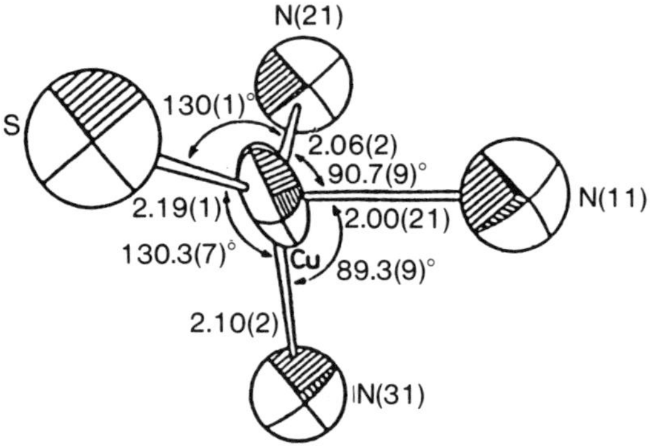

Fig. 6.8 – View of the co-ordination geometry about the copper atom in complexes [CuN_3(SR)]. [Reproduced by permission from *Proc. Natl. Acad. Sci., U.S.A.*, **74**, 3114 (1977)].

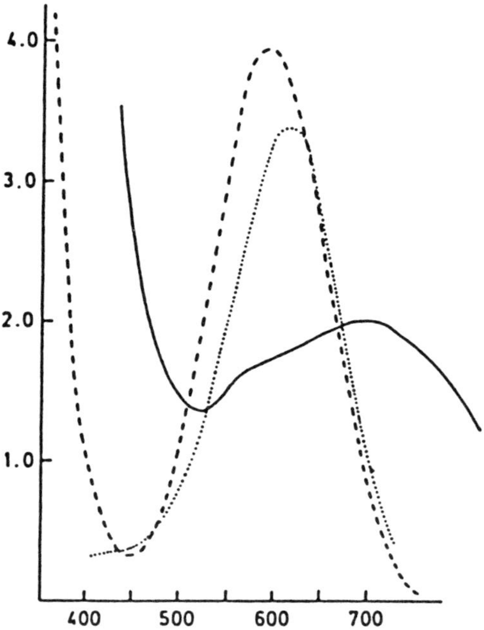

Fig. 6.9 – Optical spectra of: (– – –) [Cu-HB(3,5-Me_2pz)$_3${(*p*-$NO_2C_6H_4S$}] at 78 °C in tetrahydrofuran(——) [Cu{HB(3.5-Me_2pz)$_3$} (*O*-ethylcysteinate)] at –78 °C in tetrahydofuran; (. . . .) *Pseudomonas aeruginosa* azurin. [Reproduced by permission from *Proct. natl. Acad. Sci., U.S.A.*, **74**, 3114 (1977).

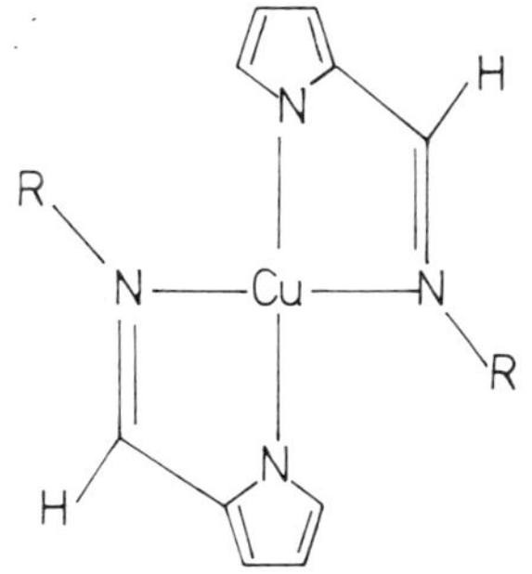

(6.10)

COPPER(II) COMPLEXES OF SMALL PEPTIDES

When copper(II) is coordinated to oxygen and nitrogen ligands the geometry is usually tetragonal with four short and two long bonds (a result of Jahn-Teller distortion). One or both of the long bonds may be removed completely, or be outside the distance considered as a bond, barely forming a Van der Waals contact. The resulting square planar and square-pyramidal geometries may be considered as a limiting case of a distorted octahedron. An empirical correlation between the stereochemistry of the copper(II) complexes of peptides and the ligand field effects produced by the *four closest donor atoms* has been noted by Freeman. The six-coordinate complexes are blue-green (λ_{max} *ca* 730 nm), fiv-coordinate species are blue (λ_{max} *ca* 635 nm) and four-coordinate complexes violet-pink (λ_{max} *ca* 500 nm). Square-pyramidal copper occurs in the copper(II) complexes of glycylglycine (**6.11**) and glycylglycylglycine (**6.12**) which have λ_{max} 635 nm and 555 nm respectively. In these structures the copper(II) ion is positioned slightly above the square plane in the direction of the fifth ligand.

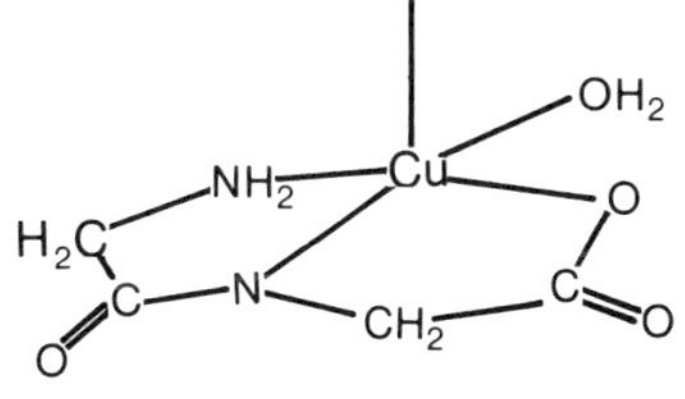

$Cu^{II}(GlyGly) . 3H_2O$

(6.11)

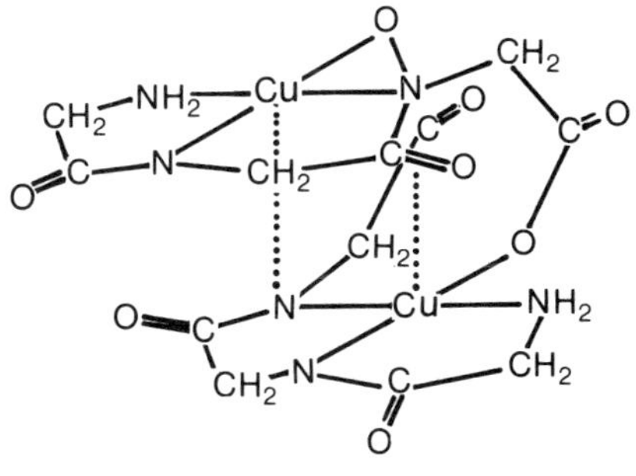

$NaCu^{II}(GlyGlyGly)H_2O$

(6.12)

Six-coordinate copper(II), which is tetragonally distorted, occurs with glycylhistidine and glycylhistidylglycine (**6.13**). $Na_2[Cu^{II}(GlyGlyGlyGly)]$ $4H_2O$ with three deprotonated amide groups has λ_{max} 510 nm and has the square planar structure (**6.14**).

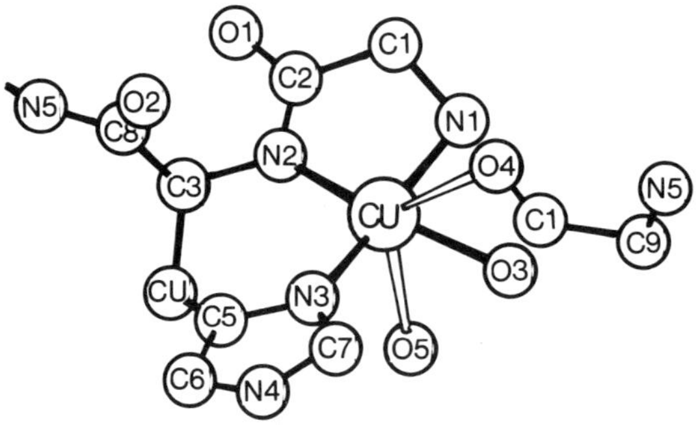

(6.13)

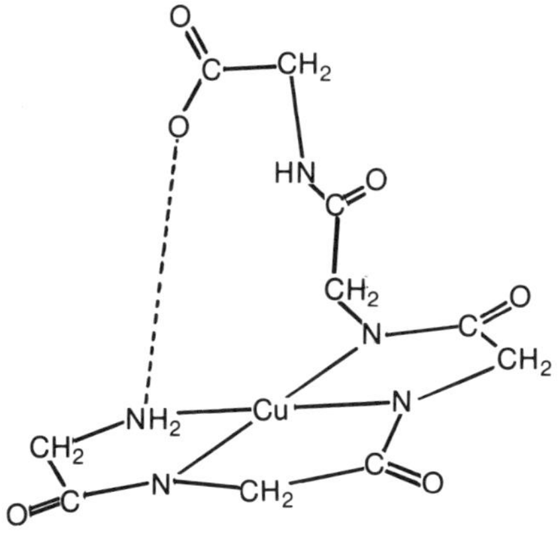

Na$_2$Cu^{II}(GlyGlyGlyGlyGly) . 4H$_2$O

(6.14)

A most unusual structure was reported for the copper(II) complex of 6-aminohexanoic acid which has the $CuO_4O'_4$ chromophore, with four short and four long bands forming a distorted dodecahedron (**6.15**).

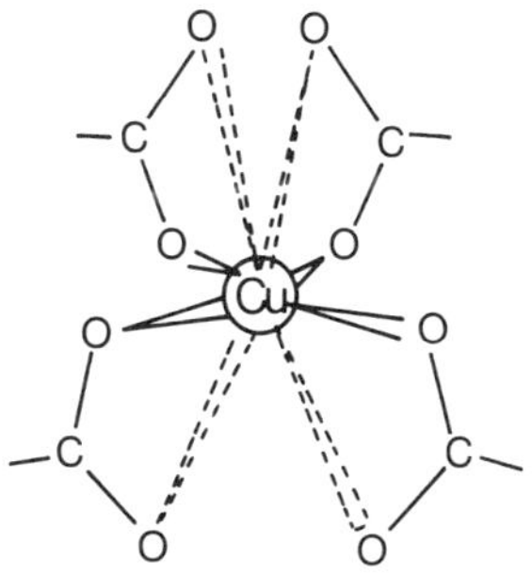

Cu^{II}(6-aminohexanoic acid)$_4$. $(ClO_4)_2$

(6.15)

Recently imidazolate-bridged copper(II) complexes have been prepared (**6.16**) with glycylglycine. The room temperature magnetic moment of (**6.16**) is 1.80 BM at 22 °C and the temperature dependence of the magnetic susceptibility reveals antiferromagnetic coupling between the two copper(II) ions. The dimeric structure is maintained even in 3×10^{-2} M aqueous solutions. It is noteworthy that the Cu-Im-Zn unit occurs in bovine superoxide dismutase, and the Fe-Im-Cu unit in cytochrome *c* oxidase (Im = deprotonated imidazole).

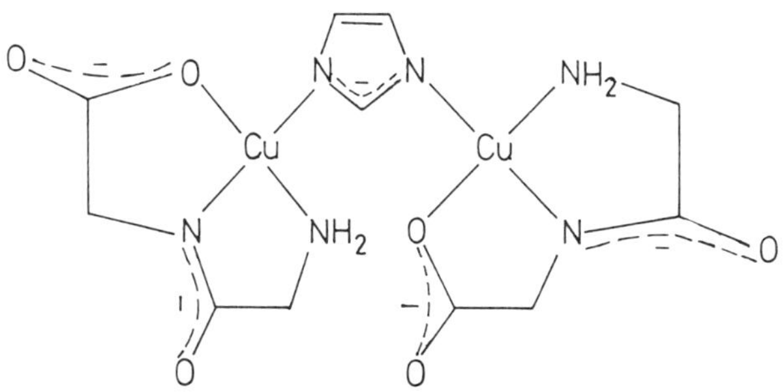

(6.16)

SPECTRAL CHARACTERISTICS

Billo has correlated spectral data for simple copper(II) peptide complexes with the type of donor groups. The wavelength of the absorbance band observed between 500 and 740 nm varies with the number of deprotonated peptide groups and amine groups. The ν_{max} ($kK = 1000\ cm^{-1}$) of the d-d band (aqueous solution spectra) can be expressed as the sum of the individual ligand field contributions of the four donor atoms which define the square plane with copper.

In equation (**6.4**) the n's refer to the number of each type of donor ($\Sigma_n = 4$)

$$\nu_{obs} = n_a\nu_N(\text{peptide}) + n_b\nu_N(\text{amino}) + n_c\nu_N(\text{imidazole}) + n_d\nu_O(\text{carboxylate}) + n_e\nu_O(\text{peptide, } H_2O \text{ or } OH^-) \qquad (6.4)$$

and ν_N(peptide) = 4.85, ν_N(amino) = 4.53, ν_N(imidazole) = 4.3, ν_O(carboxylate) = 3.42 and ν_O(peptide, H_2O or OH^-) = 3.01. The effect of axial coordination by hydroxide, carboxylate or amino groups is to shift ν_{max} to lower energy by 1 kK. Thus for a complex of the structure (**6.17**) formed between glycylglycine

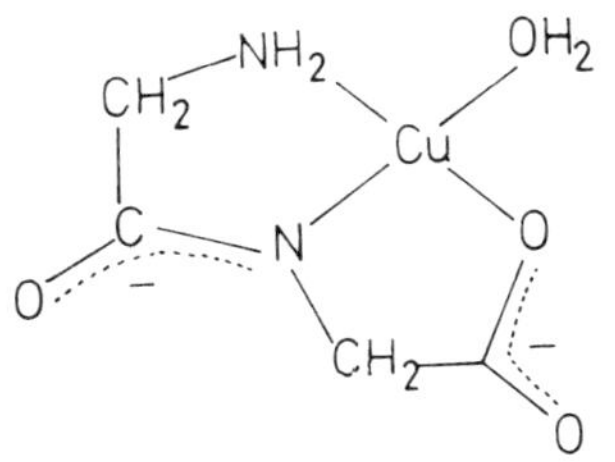

(6.17)

and copper(II) at pH 5–6, ν_{obs} = (4.53 + 4.85 + 3.42 + 3.01) = 15.81 kK = 632 nm. The spectral data for this type of complex with a selection of dipeptides is shown in Table 6.9, the observed and calculated values are in good agreement.

Table 6.9 – Absorption spectra of Cu(II)-peptide complexes.

Peptide	λ_{max}(nm)	$\epsilon(M^{-1}\,cm^{-1})$
Glycylglycine	635	84
Prolylglycine	635	100
Glycylvaline	640	95
Valylglycine	635	92

VITAMIN B_{12} AND B_{12} COENZYMES

Interest in vitamin B_{12} was aroused by the observation that it was effective in the treatment of pernicious anaemia. The structure of vitamin B_{12} is shown in Fig. 6.10. The compound contains a 15-membered corrin ring rather than a 16-membered porphyrin ring and is thus an example of a corrinoid. The cobalt atom is in the cobalt(III) oxidation state, and the two axial positions are occupied by a variety of ligands. The parent vitamin B_{12} has a benzimidazole group in one axial site and cyanide in the other. The vitamin is often called by its trivial name cyanocobalamin. An aquo derivative with H_2O replacing CN^- in the axial site is called aquocobalamin or B_{12a}. Derivatives in which the base-sugar-phosphate moiety have been removed by hydrolysis are called cobinamides. In the coenzyme forms of B_{12} which are the active species in biological systems, the cyanide ligand is replaced by an alkyl group. In the adenosyl coenzyme,

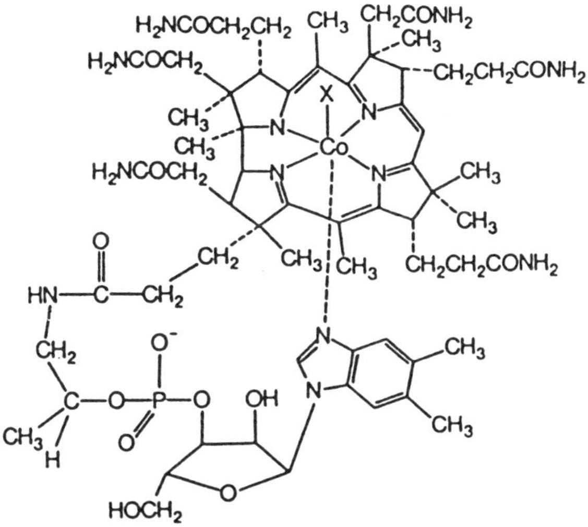

cobalamin, benzimidazole nucleotide is present
cobinamide. benzimidazole nucleotide is absent
B_{12}, X = CN, Co(III)
B_{12r} Co(II)
B_{12s} Co(I)
aquocobalamin, X = H_2O

B_{12} coenzyme, X =

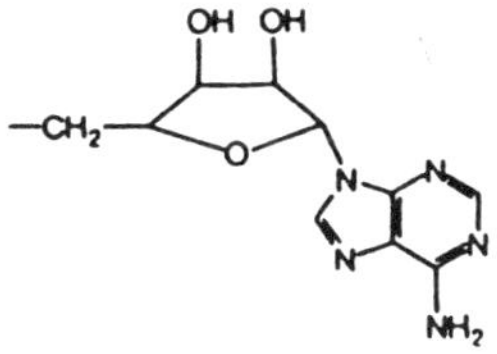

cobaloxime

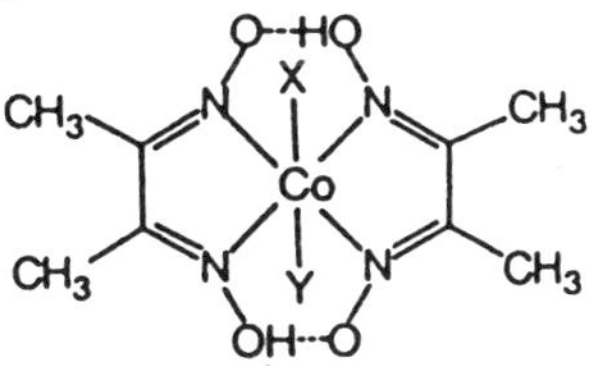

Fig. 6.10 – Structural glossary for vitamin B_{12} derivatives and models.

which is usually referred to as the B_{12} coenzyme, the adenosyl group is bound to cobalt via a cobalt-carbon bond (Fig. 6.11). These organo-cobalt complexes can be formally considered as cobalt(III) complexes containing a coordinated carbanion. More than 80% of the vitamin B_{12} present in avian and mammalian liver occurs in the coenzyme form. The methyl derivative of B_{12} which has a σ-bonded methyl group in the sixth axial site has been isolated from some microorganisms. Abbreviated structures of vitamin B_{12} derivatives are shown in Fig. 6.12.

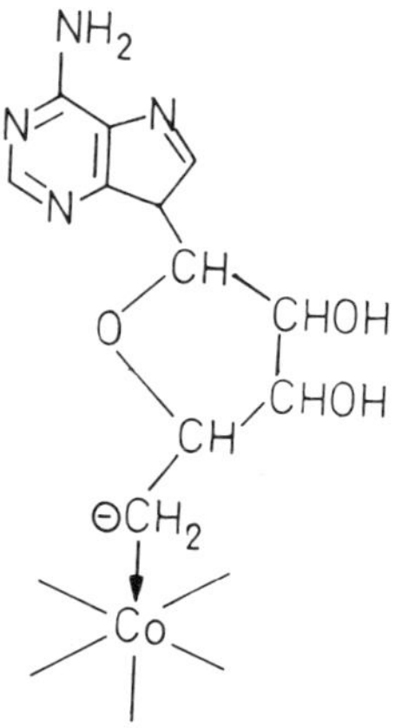

Fig. 6.11 – The 5′-deoxyadenosyl ligand present in the corrinoid 'coenzymes'.

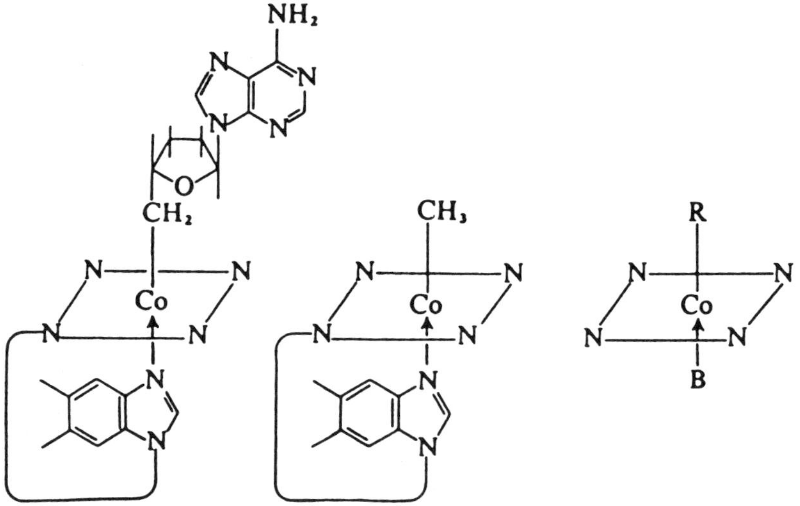

Fig. 6.12 – The structures of vitamin B_{12} derivatives: (a) B_{12} coenzyme; (b) methyl-B_{12}; and (c) alkyl derivative B_{12}.

There are two one-electron reduction steps of B_{12}, from B_{12a} (i.e. aquo) to $B_{12r}(Co^{II})$ and from B_{12r} to B_{12s} (Co^{I}). Vitamin B_{12r} has been shown to be a low-spin cobalt(II) complex (d^7) and can be obtained by chemical or electrolytic reduction of B_{12a}. The e.p.r. spectra fall into two different groups suggesting four-coordinate (planar) complexes and five-coordinate complexes. It appears that both types of complex have the same electronic configuration $(d_{yz})^2(d_{zx})^2$ $(d_{x^2-y^2})^2(d_{z^2})^1(d_{xy})^0$. The odd electron in the d_{z^2} orbital in B_{12r} would be expected to behave like a free radical and it combines reversibly with dioxygen to give a mononuclear dioxygen complex.

$B_{12s}(Co^{I})$ can be produced from B_{12} or B_{12r} (Co^{II}) by enzymic reduction, by chemical reduction with Na/Hg or BH_4^-, and by electrolytic reduction. Cobalt(I) has a d^8 configuration and is usually low spin. B_{12s} and cobalt(I) model compounds have either four-coordinate (planar) or five-coordinate (tetragonal pyramid) structures with a $(d_{yz})^2(d_{zx})^2(d_{x^2-y^2})^2(d_z)^2(d_{xy})^0$ electronic structure. Due to the low oxidation state and tetragonal structure, the $(d_{z^2})^2$ electrons behave as a strongly nucleophilic lone pair. The nucleophilicity of Co(I) in B_{12s} and in model compounds has been studied in detail by Schrauzer and his co-workers. Their results have been expressed in terms of n-values which provide a measure of the nucleophilicity of the nucleophile towards methyl iodide. The n-value is the logarithm of the rate constant for the reaction of the nucleophile with methyl iodide, for example.

$$Co^{I}LB + CH_3I \xrightarrow{k} \underset{(Co^{III})}{CH_3\text{-}CoLB} + I^-$$

Where $n = \log k$. Typical values of n are listed in Table 6.10. The very high nucleophilicity of Co(I) in B_{12s} and in model complexes of dimethylglyoxime-(dmg) are apparent.

Table 6.10 – Nucleophilicity of $Co^{I}(B_{12s})$, $Co^{I}(dmg)_2B$, and other model compounds.

Nucleophile	n_{CH_3I}	Nucleophile	n_{CH_3I}
CH_3OH	0.00	$Co^{I}(dmg)_2(P\text{-}nC_4H_9)_3$	13.3
NH_3	5.50	$Co^{I}(dmg)_2(py)$	13.8
I^-	7.42	$Co^{I}(B_{12s})$	14.4
$P(n\text{-}C_4H_9)_3$	8.69		

Source: G. N. Schrauzer, E. Deutsch and R. J. Windgassen, *J. Am. Chem. Soc.*, **90**, 2441 (1968). Reprinted with permission of the American Chemical Society. Copyright by the American Chemical Society.

ENZYMIC REACTIONS

The main corrinoids in human serum and tissues are hydroxocobalamin (OH-Cbl) methylcobalamin (Me-Cbl) and adenosyl-cobalamin(Ado-Cbl). Ado-Cbl is a cofactor for the conversion of methylmalonyl-CoA into succinyl-CoA (equation 6.5), catalysed by methylmalonyl CoA mutase (SCoA = coenzyme A). Me-Cbl participates in the synthesis of methionine **(6.19)** from homocysteine **(6.18)**.

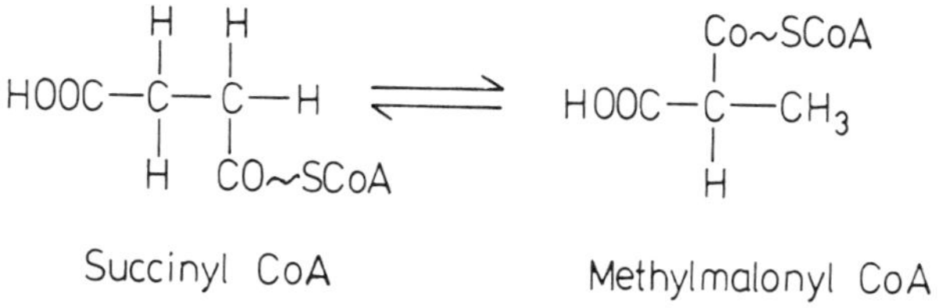

(6.5)

$$\underset{\text{Homocysteine }(\mathbf{6.18})}{\underset{\displaystyle NH_2}{HSCH_2\overset{}{C}HCO_2H}} \longrightarrow \underset{\text{Methionine }(\mathbf{6.19})}{\underset{\displaystyle NH_2}{CH_3SCH_2CH_2\overset{}{C}HCO_2H}}$$

Methionine synthetase catalyses the transfer of a methyl group from N^5-methyl-tetrahydrofolate (Me-THF) to homocysteine to form methionine. It is believed that the corrinoid plays the role of transferring the methyl group via the formation of methyl-B_{12}. The scheme in Fig. 6.13 has been proposed for this reaction

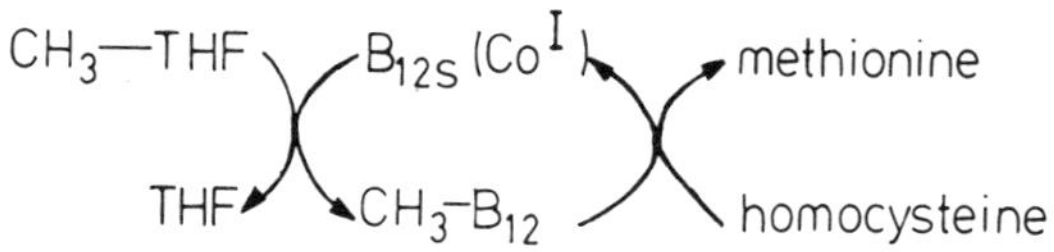

Fig. 6.13 – Action of methionine synthetase.

The formation of $CH_3 - B_{12}$ from the reaction of B_{12s} with CH_3-THF and the subsequent transfer of the methyl group on the cobalt to homocysteine has been established by ^{14}C tracer experiments.

No cobalamin-dependent enzyme has been sequenced and only one has been crystallized (methylmalonyl-CoA mutase from *Propionbacterium sherimanii*, as a complex with OH – Cbl). Mechanistic discussions have therefore been directed to understanding the function of the cobalamin, whereas the protein has been of necessity ignored.

Enzymes containing cobalt corrinoids catalyse three types of reaction:

(i) The transfer of methyl groups, for example in the formation of methionine from homocysteine and in the formation of methyl mercury compounds equation (6.6):

$$\underset{(Co^{III})}{CH_3{-}B_{12}} + HgCl_2 \longrightarrow CH_3HgCl + \underset{(Co^{I})}{B_{12s}} + Cl^- \qquad (6.6)$$

(ii) The so-called 'isomerase' reaction:

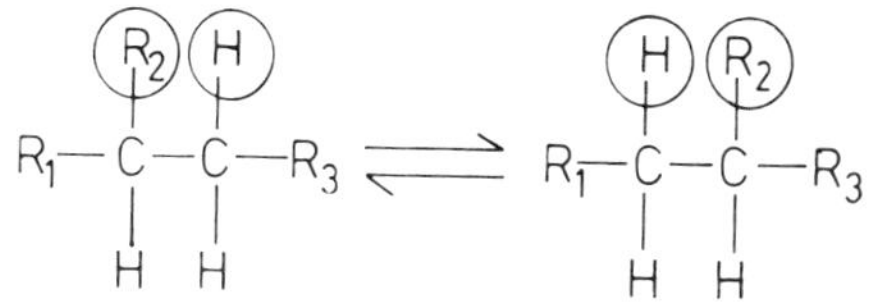

which involves the 1,2-shift of a C, N, or O atom. Typical 'isomerase' reactions are listed in Table 6.11. The interconversion of succinylcoenzyme A and L-methylmalonylcoenzyme A fall into this category.

(iii) The reduction of the $-CHOH-$ group of ribonucleotide triphosphates to $-CH_2-$.

Reactions of group(i) involve the intermediate formation of a Co^{III}-CH_3 complex while the reactions of types (ii) and (iii) both require the complex to be present in the 'coenzyme' form which contains the 5′-deoxyadenosyl ligand. The mechanism of the 'isomerase' type of reaction is still very controversial and different mechanisms has been proposed by Abeles, Schrauzer, Corey and their coworkers. The interested reader may consult reference [21] for a detailed discussion of this problem.

MODEL COMPOUNDS

One of the remarkable properties of B_{12} is its ability to form alkyl derivatives, including adenosyl-B_{12} (the coenzyme). The bond is mainly of a σ-type between the cobalt atom and the carbon atom of the alkyl group and these compounds fall within the classification of organometallic compounds. The σ-bonded alkyl derivatives of transition metals, particularly those of the first transition series, have until recently been regarded as being somewhat unstable unless the transition metals are in low oxidation states and a π-acid ligand such as CO is present. It is now recognized that these compounds may be kinetically unstable in view of the possible β-elimination pathway for their decomposition (equation (6.7)).

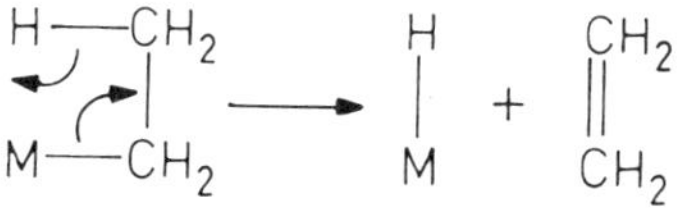

(6.7)

Table 6.11 – 'Isomerase' reactions requiring 5′-deoxyadenosylcorrinoids.

Substrate	R_1	R_2	R_3	Product
			irreversible reactions	
Ethane-1,2-diol	H	OH	OH	acetaldehyde
Propane-1,2-diol	CH_3	OH	OH	propionaldehyde
Glycerol	$HOCH_2$	OH	OH	β-hydroxypropionaldehyde
Ethanolamine	H	NH_2	OH	acetaldehyde + NH_3
			reversible reactions	
L-Glutamate	H	$-CHNH_2 . COOH$	COOH	*threo*-β-methylaspartate
Succinyl-coenzyme A	H	—CO . SR	COOH	L-methylmalonylcoenzyme A
α-methyleneglutarate	H	$-C(=CH_2) . COOH$	COOH	β-methylitaconate
Ornithine	H	NH_2	$-CH_2 . CHNH_2 . COOH$	2,4-diaminovalerate
L-β-Lysine	H	NH_2	$-CH_2 . CHNH_2 . CH_2 . COOH$	3,5-diaminohexanoate
D-α-Lysine	H	NH_2	$-CH_2 . CH_2 . CHNH_2 . COOH$	2,5-diaminohexanoate

Quite a large variety of σ-bonded alkyl derivatives of cobalt(III) have now been prepared. These complexes are formally regarded as cobalt(III) complexes with R^- carbanion donors. Some of these compounds are shown in Fig. 6.14.

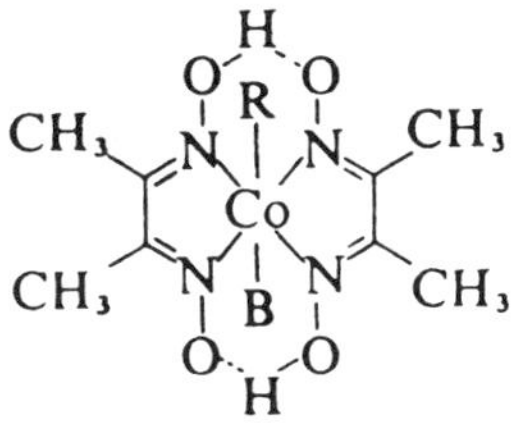

$[RCo(dmg)_2B]$
(a)

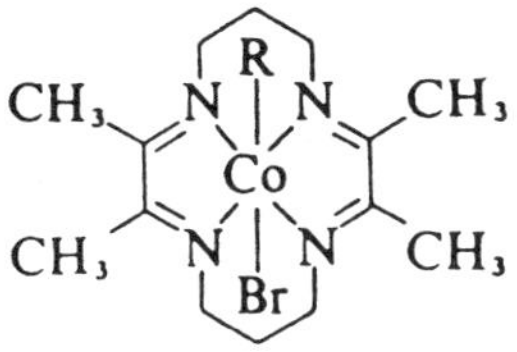

$[RCo(tim)Br]^+$
(b)

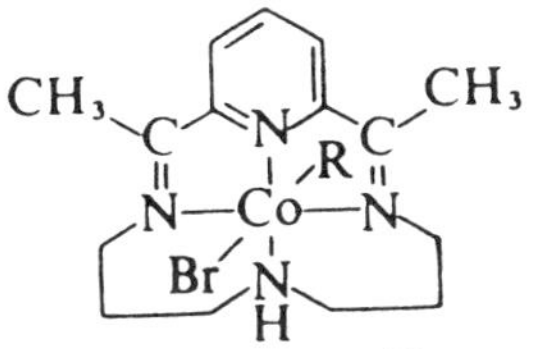

$[RCo(cr)Br]^+$
(c)

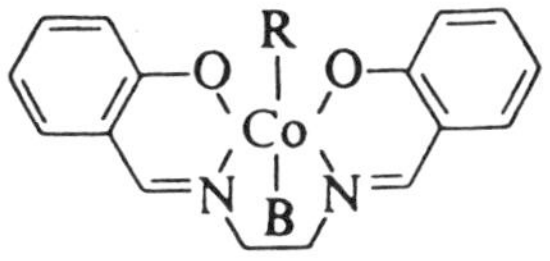

[RCo(salen)B]
(e)

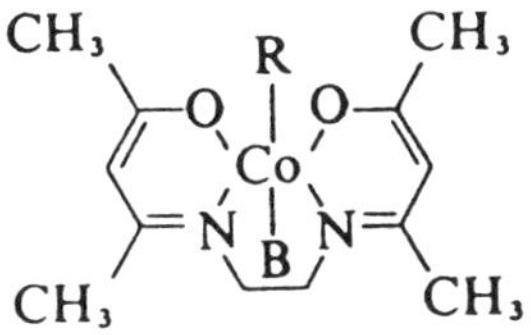

[RCo(acacen)B]
(d)

Fig. 6.14 – The structures of the B_{12} model compounds. (a) Schrauzer, *Accounts Chem. Res.*, **1**, 97 (1968). (b) Farmery and Busch, *Chem. Commun.*, 1970, 1091 (1970). (c) Ochiai and Busch, *Chem. Commun.*, 1968, 905 (1968). (d) Costa *et al.*, *J. Organomet. Chem.*, **6**, 181 (1966). (e) Costa *et al.*, *J. Organomet. Chem.*, **11**, 333 (1968).].

The cobaloximes (i.e. from 'cobalt-oxime') prepared using the dimethylglyoxime ligand (dmgH) have been used as models for many B_{12} reactions. All of these complexes have a fair degree of π-conjugation. A porphyrin has an extensive π-conjugated system of 24 electrons (it is in fact an 18-π system) whereas a corrin is less π-conjugated with 13 electrons. The cobaloxime and TIM ligands have a 4×2 π-electrons conjugated, acacen has 5×2, salen has 9×2 and CR has 10. A cobalt complex requires some π-conjugation in the planar ring in order to form a σ-bonded alkyl derivative. However, too much conjugation does not appear to be favourable. Ligands such as TIM in fact stabilize dialkyl derivatives which are not known for the B_{12} system, and although porphyrins will form alkyl derivatives, they are not formed readily. The fact that the corrin ring has extensive but not excessive π-conjugation is of importance in the B_{12} reactions.

REFERENCES AND BIBLIOGRAPHY

Cytochrome c

[1] E. Margolaish and A. Schejter, *Adv. Protein Chem.,* **21**, 113 (1966).
[2] R. E. Dickerson and R. Timkovich, in *The Enzymes,* ed. P. D. Boyer, 1975, Vol. XI, p. 397.
[3] F. R. Salemme, *Ann. Rev. Biochem.,* **46**, 299 (1977).
[4] E. Margoliash, *Adv. Chem. Phys.,* **29**, 191 (1977).

Superoxide Dismutase and Superoxide

[5] I. Fridovich, *Acc. Chem. Res.,* **5**, 321 (1972).
[6] I. Fridovich, *Ann. Rev. Biochem.,* **44**, 147 (1975).
[7] U. Weser, *Struct. Bonding,* **17**, 1 (1973).
[8] A. M. Michelsen and J. M. McCord, *Superoxide and Superoxide Dismutases,* Academic Press, 1977.
[9] O. Hayaishi and K. Asada eds., *Biochemical and Medical Aspects of Active Oxygen,* University of Tokyo Press, 1977.
[10] I. Fridovich, Oxygen radicals, hydrogen peroxide and oxygen toxicity, in *Free Radicals in Biology,* ed. W. A. Pryor, Academic Press, 239-277, 1976.
[11] O. Haiyashi and K. Asada, eds., *Biochemical and Medical Aspects of Active Oxygen,* University of Tokyo Press, 1977.
[12] S. J. Lippard, A. R. Burger, K. Ugurbil, J. S. Valentine and M. W. Pantoliano, 'Physical and Chemical Studies of Bovine Erythrocyte Superoxide Dismutase, in *Bioinorganic Chemistry II, (Adv. Chem. Ser.,* 162), American Chemical Society, Washington, 1977, p. 251-262.

Blue-Copper Proteins

[13] E. L. Ulrich and J. L. Markley, 'Blue-Copper Proteins. Nuclear Magnetic Resonance Investigations, *Coord. Chem. Rev.,* **27**, 109 (1978).
[14] P. M. Collman, H. C. Freeman, J. M. Guss, M. Murata, V. A. Norro, J. A. M. Ramshaw and M. P. Venkatappa, *Nature,* **272**, 319 (1978). X-Ray work on plastocyanin).

Cytochrome P-450

[15] R. Sato and T. Omura, eds., *Cytochrome p-450,* Academic Press, New York, 1978.
[16] R. E. White and M. J. Coon, *Ann. Rev. Biochem.,* **49**, 315 (1980).

[17] L. S. Alexander and H. M. Goff, 'Chemicals, Cancer and Cytochrome p-450' *J. Chem. Ed.*, **59**, 179 (1980).

[18] J. P. Collman and S. E. Groh, 'Mercaptan-Tail' Porphyrins: Synthetic Analogues for the Active Site of Cytochrome p-450, *J. Am. Chem. Soc.*, **104**, 1391 (1982).

Vitamin B_{12}

[19] J. M. Pratt, *Inorganic Chemistry of Vitamin B_{12}*, Academic Press, London, 1972.

[20] H. A. O. Hill, J. M. Pratt and R. J. P. Williams, The Chemistry of Vitamin B_{12}, *Chem. Brit.*, 156 (1969).

[21] For a review of cobalamin chemistry and biochemistry, see B. T. Golding in *Comprehensive Organic Chemistry*, ed. W. Haslam, Pergamon Press 1978, Vol. 5, Chapter 24.4.

CHAPTER 7

Nitrogen Fixation and Iron-Sulphur Proteins

INTRODUCTION

Biological nitrogen fixation is the process whereby some bacteria and blue-green algae convert atmospheric nitrogen into ammonia. Estimates (1974) suggest that approximately 175 million metric tons of nitrogen are fixed annually in this way with 50 million metric tons via the Haber industrial process.

Molecular nitrogen, or dinitrogen (N_2) is the most inert diatomic molecule. The molecule owes its lack of reactivity to the large energy difference between its filled and vacant molecular orbitals. The filled MOs are low in energy ($\leqslant -15.6$ ev) and its vacant orbitals are high ($\geqslant -7$ ev). As a result it is very difficult to add electrons to the dinitrogen molecule, or to remove them from it in the ground state. Hence N_2 at room temperature is inert to strong oxidizing agents such as F_2, and strong reducing agents such as sodium metal. Some electropositive metals such as lithium, calcium, magnesium and titanium react with N_2 at or above room temperature to form stable nitrides:

$$3\,Mg + N_2 \longrightarrow Mg_3N_2$$

Industrially NH_3 is synthesized by the Haber process using moderate pressures and a reduced iron catalyst:

$$3\,H_2 + N_2 \longrightarrow 2NH_3$$

NITROGENASE

The enzyme system responsible for fixing N_2 is known as nitrogenase. Two protein components, the Fe protein and the Mo-Fe protein, make up the nitrogenase enzyme complex. The major elements of the nitrogenase reaction are summarized in Fig. 7.1. Acetylene is also reduced to C_2H_4 by the enzyme (Fd = ferredoxin). Electrons flow from a reducing agent (Fd_{red}) into the Fe protein, then into the Mo-Fe protein and finally onto the substrate

$$N_2 + 6H^+ + 6e \longrightarrow 2NH_3$$

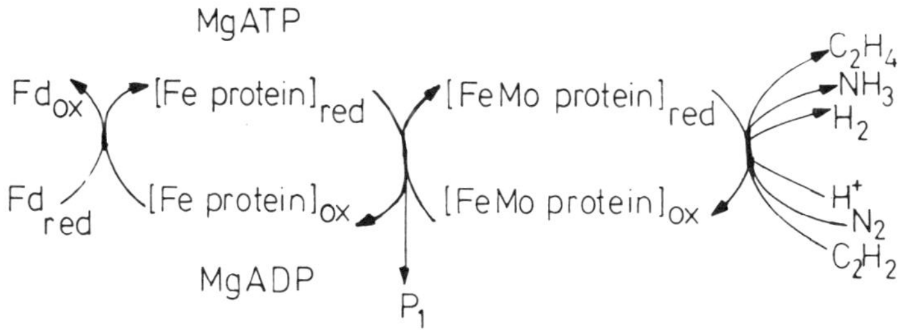

Fig. 7.1 – The nitrogenase reaction.

MgATP which binds to the Fe protein is hydrolysed as the substrate is reduced. The precise time at which MgATP binds and dissociates, and the exact role it plays in the reduction process has not been fully elucidated.

Chromatography of nitrogenase on a Sephedex column separates the Fe protein and the Mo–Fe protein. The Mo–Fe protein is brown (M = 200,000–220,000) which ideally contains two atoms of Mo per molecule, but more usually 1.3–1.8 atoms. The protein also contains about 24 atoms of iron, but 18 to 34 atoms have been reported.

The Fe protein is yellow (M = 65,000) and contains four atoms of Fe, but no molybdenum. Both proteins appear to contain sulphide ions equal in number to the Fe atoms. The iron is present mainly as Fe_4S_4 cubane clusters, which provide a system for electron storage. Both proteins are air sensitive. The Mo–Fe protein will withstand brief exposure to air, and can then be reduced again to the active form. The Fe protein is destroyed irreversibly by the briefest exposure to air.

Together the two proteins will catalyse the reduction of N_2 to NH_3 in neutral aqueous solution in the presence of MgATP and a suitable reducing agent such as dithionite. Twelve molecules of ATP are hydrolysed to ADP and orthophosphate, in the reduction of one molecule of N_2 to ammonia. Hence, the natural process is very energy intensive ($\Delta G^{0\prime}$ for ATP hydrolysis is – 30.5 kJ mol^{-1}, see Chapter 9).

The Fe protein acts as a specific electron carrier to the Mo–Fe protein, with which it forms a complex. The electron transfer from the Fe to the Mo–Fe protein only occurs in the presence of MgATP. Currently it is thought that the N_2 molecule attaches itself to one or both molybdenum atoms in the enzyme, and is protonated to NH_3 with a simultaneous uptake of electrons via the Mo atom from the Fe-sulphur electron storage system.

DINITROGEN COMPLEXES

The first dinitrogen complex to be characterized, $[Ru(NH_3)_5N_2]^{2+}$, was prepared in 1965 by Allen and Senoff. Over 100 dinitrogen complexes have now been characterized. Many of the early complexes were prepared accidentally but clear-cut synthetic routes are now available. These routes include (a) direct reaction with N_2, by replacement of a labile ligand or by reduction of a transition metal complex in the presence of N_2 (b) the reaction of certain coordinated ligands, for example the oxidation of hydrazine, and (c) metathesis.

Dinitrogen is usually bound end-on to transition metals (**7.1**), with a N-N

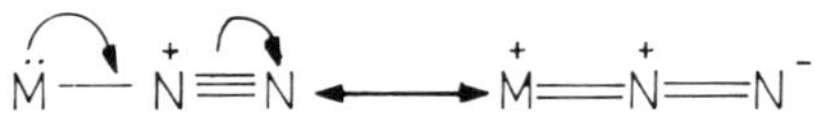

(7.1)

bond length of 1.123–1.124 Å, somewhat longer than the N-N bond length of 1.097 Å in free N_2. This lengthening of the N-N bond arises due to back donation of d-electron density into the π^* orbitals of N_2 (Fig. 7.2).

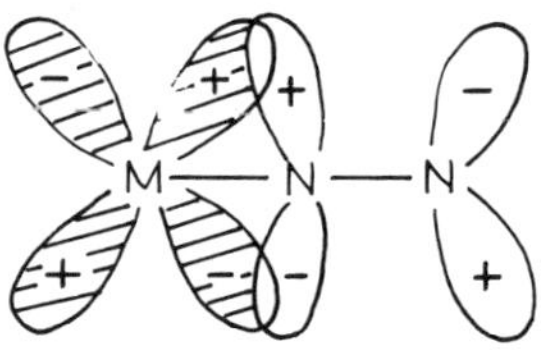

Fig. 7.2 – $d\pi \rightarrow p\pi^*$ bonding in N_2 complexes.

A few compounds are known in which N_2 is bound side on (both in terminal and bridging stituations). The complex $[RhCl(N_2)(PPr^i_3)_2]$ has a terminal side-on ligand and the N-N bond length is only slightly longer than in N_2.

The extent of back donation from the metal d-orbitals into the π^* orbitals of N_2 is indicated by the ν_{N_2} frequency and by the intensity of the absorption. A strong $d\pi \rightarrow p\pi^*$ interaction is indicated by a low ν_{N_2} and high intensity. The ν_{N_2} (Raman active) band in N_2 occurs at 2331 cm^{-1} and at 2105-2167 cm^{-1} in the complex $[Ru(NH_3)_5N_2]X_2$ depending on the counter ion X^-.

THE REACTIVITY OF COORDINATED N_2

Attempts to reduce coordinated N_2 in transition metal complexes met with little initial success. The first demonstration of partial reduction involved the treatment of the phosphine complex $[M(N_2)_2(Ph_2PCH_2CH_2PPh_2)_2]$ (M = Mo or W) with HBr leading to coordinated diimine N_2H_2:

$$[Mo(N_2)_2(diphos)_2] \xrightarrow{HBr} [MoBr_2(N_2H_2)(diphos)_2] + N_2$$

Further experimental work indicated that complexes $[M(N_2)_2(PR_3)_4]$ (M = Mo or W) containing tertiary phosphines, on treatment with $MeOH/H_2SO_4$ gave almost complete conversion of 1 mole of N_2 to NH_3, with a little hydrazine NH_2NH_2. A 90% yield of NH_3 was obtained with *cis*-$[W(N_2)_2(PMe_2Ph)_4]$. The following scheme has been suggested for the stepwise reduction:

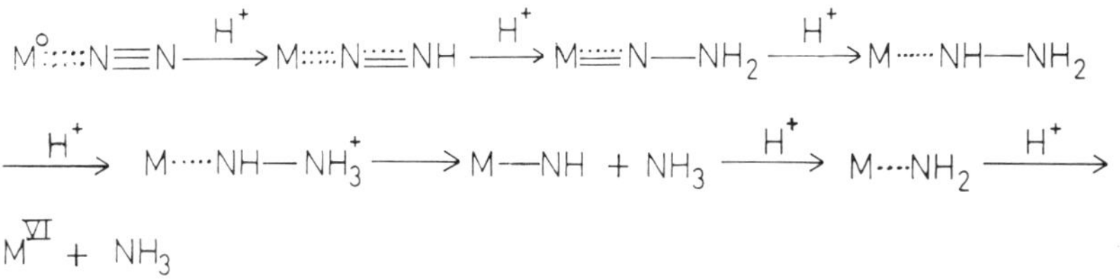

The M^0 atom provides the six electrons for the reduction $N_2 + 6H^+ + 6e \rightarrow 2NH_3$.

THE FUTURE

It is doubtful if the Haber process for the synthesis of NH_3 will be replaced in the forseeable future by a catalytic system based on N_2-metal complexes.

Probably the most important potential development, to provide additional nitrogen for crops, is to transfer the structural and regulatory genes involved in N_2 fixation, to plants such as corn which cannot fix N_2. In 1972, Postgate and coworkers were able to confer nitrogen fixing properties upon the common gut bacterium, *Escherichia coli* by transferring genes from *Klebsiella pneumoniae* bacteria. Later the *K. Pneumoniae nif* genes (nif = nitrogen fixation) which they incorporated in *E. coli* cells were moved into a plasmid which carried a sex factor F. (Plasmids are pieces of DNA which exist within a cell, but outside of the nuclear body where DNA is usually located). This new element was called *F′ nif.* The association of the *nif* genes with the sex factor allowed the genes to be transferred to a variety of bacteria so developing nitrogen-fixing *Salmonella typimurium* and *K. aerogenes.*

IRON-SULPHUR PROTEINS

Since the initial isolation of purified ferredoxin proteins from *Clostridium pasteurianum* and spinach in 1962, non-haem iron-sulphur proteins have been the subject of intense activity. Iron-sulphur proteins are widely distributed in nature, being found in anaerobic, aerobic and photosynthetic bacteria, algae, fungi, higher plants and mammals. Their biological function is in electron transfer reactions. In view of their spectroscopically responsive iron atoms, they have aroused considerable interest.

Structures

In the lower molecular weight iron–sulphur redox proteins, there are three types of active site which in terms of minimal composition are $[Fe(S\text{-}Cys)_4]$ (7.2), $[Fe_4S_4^*(S\text{-}Cys)_4]$ (7.3) and $[Fe_2S_2^*(S\text{-}Cys)_4]$ (7.4) where S* is S^{2-}. Structure (7.2) has been established by X-ray diffraction studies of oxidized rubredoxin (Rb_{ox}) (M = 6000) from *C. pasteurianum.* Fig. 7.3 illustrates the amino-acid sequence of the rubredoxin from *Micrococcus aerogenes.*

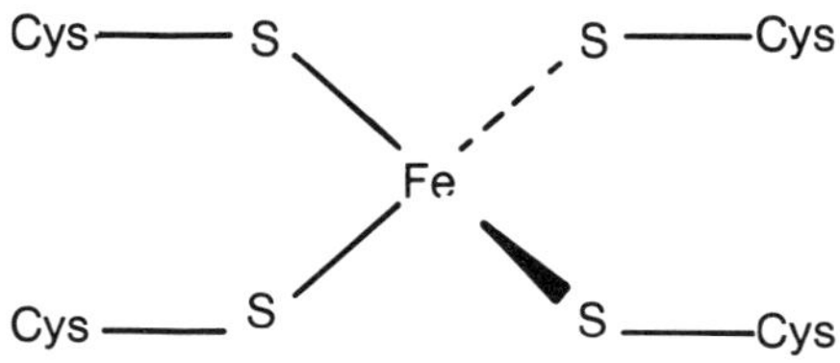

(7.2)

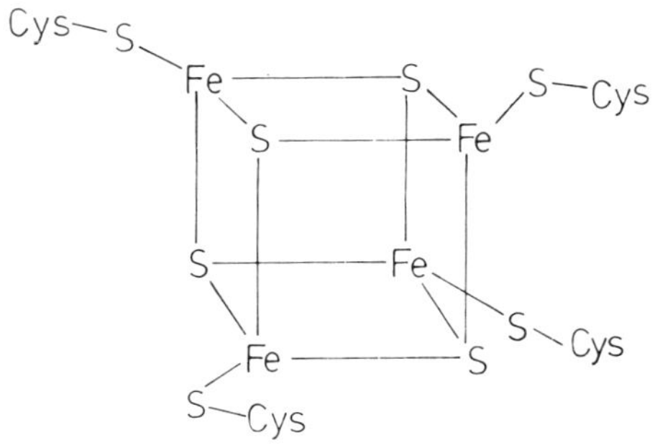

(7.3)

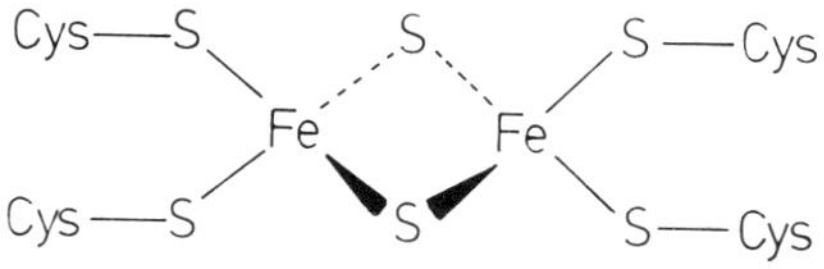

(7.4)

The structure of *C. pasteurianum* rubredioxin (oxidized) is shown in Fig. 7.4, it contains a Fe(III)–S_4 coordination unit, distorted from tetrahedral symmetry.

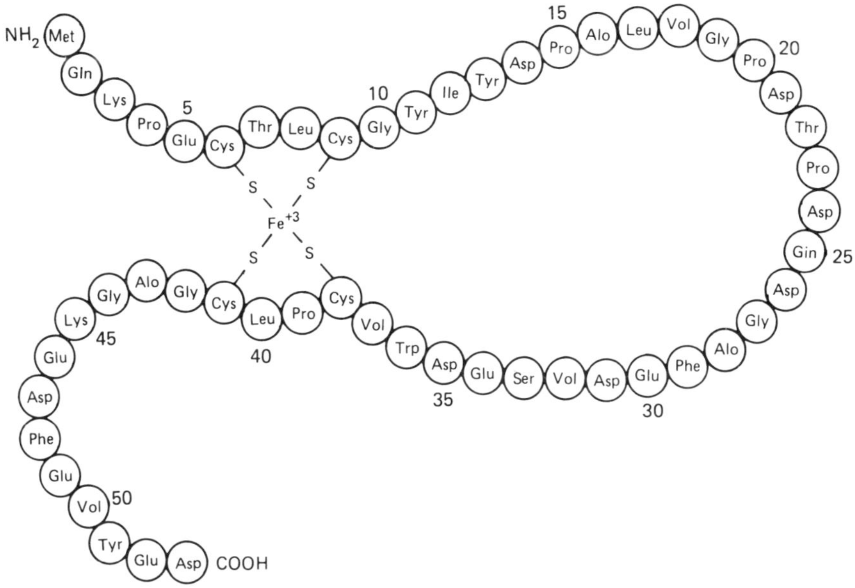

Fig. 7.3 – The amino-acid sequence of the rubredoxin from *Micrococcus aerogenes.* (R. H. Holm, *Endeavour,* **34,** 38 (1975).

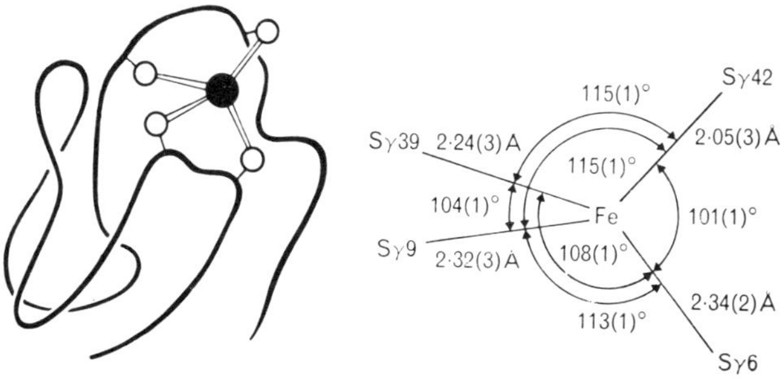

Fig. 7.4 – The structure of *Clostridium pasteurianum* rubredoxin [10, 11]. •, Fe; ○, S-Cys. (Reproduced by permission of Prof. L. H. Jensen.)

The cubane structure (**7.3**) of D_{2d} or lower symmetry has been found in the oxidized and reduced 4-Fe 'high potential' iron protein (HIPIP) with $M = 9650$ from *Chromatium,* and the oxidized 8-Fe ferredoxin (Fd_{ox}, $M = 6000$) from *Peptococcus aerogenes* (Fig. 7.5).

The binuclear structure (**7.4**) has not yet been established by X-ray methods, but is fully consistent with a large body of physiochemical evidence for oxidized and reduced 2-Fe ferredoxins. The properties of some representative 8-Fe iron-sulphur proteins are listed in Table 7.1.

Table 7.1 – Representative iron-sulphur proteins

Active group (atoms/molecule)	Molecular weight	No. of amino acids	Redox potential (E_0' in mV)
8Fe, 8S	6,000	55	−395
8Fe, 8S	6,000	60	–
8Fe, 8S	10,000	81	−490
8Fe, 8S	13,000	–	–
8Fe, 8S	15,000	130	−420

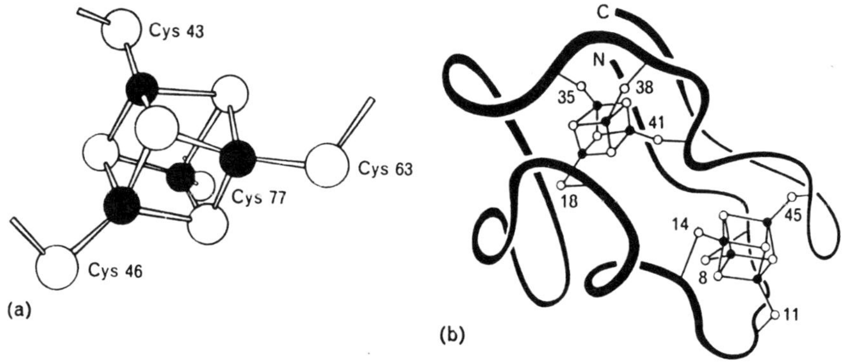

Fig. 7.5 – (a) Structure of the active site of *Chromatium* $HIPIP_{red}$. (b) Schematic structure of *P. aerogenes* Fd_{ox} showing the two 4-Fe clusters [from Jensen, L. H., Sieker, L. C., Watenpaugh, K. D., Adman, E. T. and Herriott, J. R., *Biochem. Soc. Trans.*, **1**, 27 (1973)]. •, Fe; ○, Sγ-Cys. (Reproduced by permission of the Biochemical Society and Prof. L. H. Jensen.)

A substantial number of Rd, Fd and HPIP proteins have been sequenced, and in each type of active site structure, cysteinate sulphur functions as the terminal ligand. Metal bridging is by sulphide ('acid labile' sulphur) such that sites (**7.3**) and (**7.4**) contain substructural $Fe_4S_4^*$ and $Fe_2S_2^*$ cores, the existence and stability of which lead to biologically important reactivity. In each type of site the iron atoms are coordinated in an approximately tetrahedral arrangement.

Ferredoxins

Iron-sulphur proteins catalyse oxidation-reduction reactions between +350 and −600 mV (hydrogen electrode = −420 mV at pH 7). One of the best known groups of iron-sulphur proteins are the ferredoxins (Fd). These are small proteins consisting of relatively few amino acids usually between 55 and 120

(M = 6000–14,000). They contain either 4Fe–4S or 2Fe–2S active centres which can be extracted intact from the protein in the presence of appropriate ligands and solvents ('core extrusion'). Another important aspect of removing the active centre is that the apoprotein thus formed can be reconstituted under anaerobic, non-enzymic conditions by the simple addition of iron(II) ammonium sulphate and sodium sulphide.

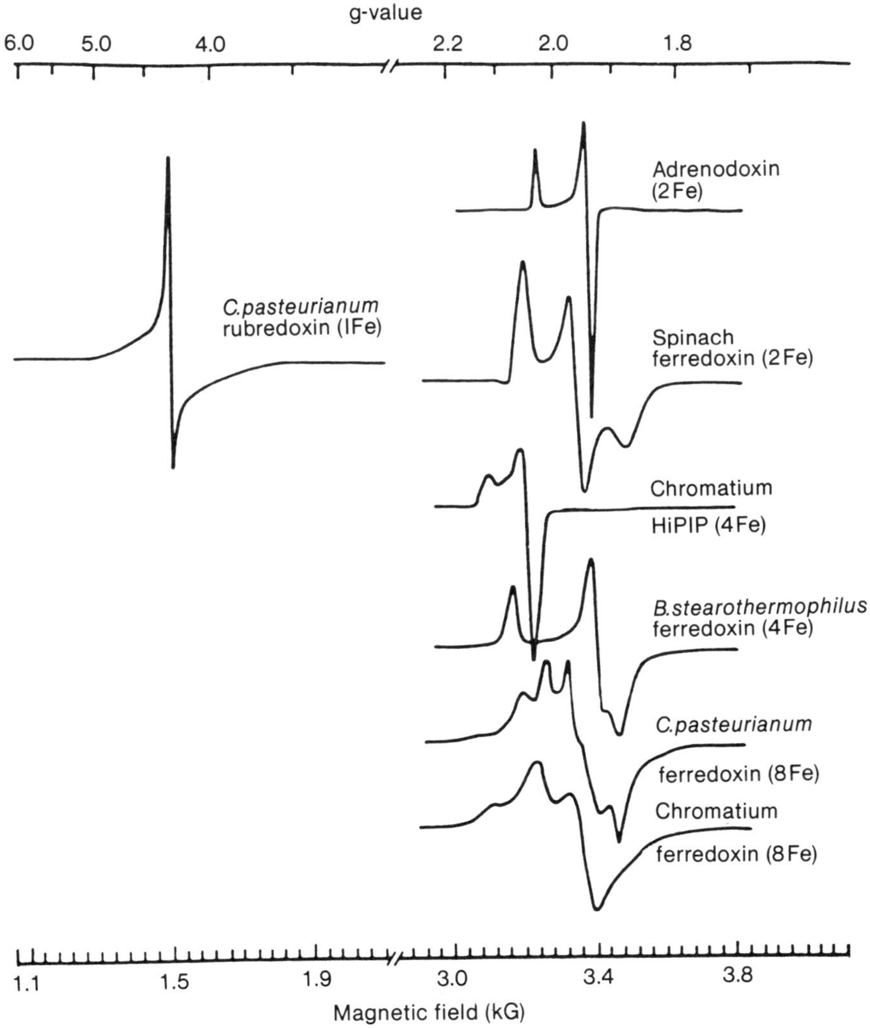

Fig. 7.6 – Representative e.s.r. spectra of iron-sulphur proteins [from Hall, D. O., Rao, K. K. and Mullinger, R. N., *Biochem. Soc. Trans.*, **3**, 472 (1975)].

The amino-acid sequence in the ferredoxins is quite distinctive, the position of the cysteines which bind the iron in a specific environment does not vary in different classes of ferredoxins. This invariance in the positions of key amino acids has been useful in studying the chemical evolution of ferredoxins. The investigations of iron–sulphur proteins have been greatly helped by the recognition of the so called $g_\perp \cong 1.94$ e.s.r. signal which is generally seen in the reduced form of iron–sulphur proteins. In the absence of e.s.r. measurements the study of iron–sulphur proteins is difficult as a result of their low extinction coefficients and broad absorption in the visible region. Figure 7.6 shows characteristic e.s.r. spectra of representative iron–sulphur proteins including the high-potential iron protein (HIPIP) which has an e.s.r. signal of $g_\perp = 2.02$ in the oxidized state.

The existence of this HIPIP-type structure was the starting point of an interesting development in the study of the iron–sulphur active centre. X-Ray crystallography had shown that there is little difference between the 4Fe–4S cluster in ferredoxin ($E^{0'} = -400$ mV) and HIPIP ($E^{0'} = +350$ mV). This anomaly was rationalized in terms of a super-reduced HIPIP and a superoxidized ferredoxin (Fig. 7.7). The super-reduced HIPIP was shown to exist by Cammack using 80% DMSO to distort the protein environment of HIPIP. In this case a super-reduced HIPIP with an e.p.r. signal similar to that of a 4Fe ferredoxin was detected. A superoxidized ferredoxin has also been reported by Sweeney.

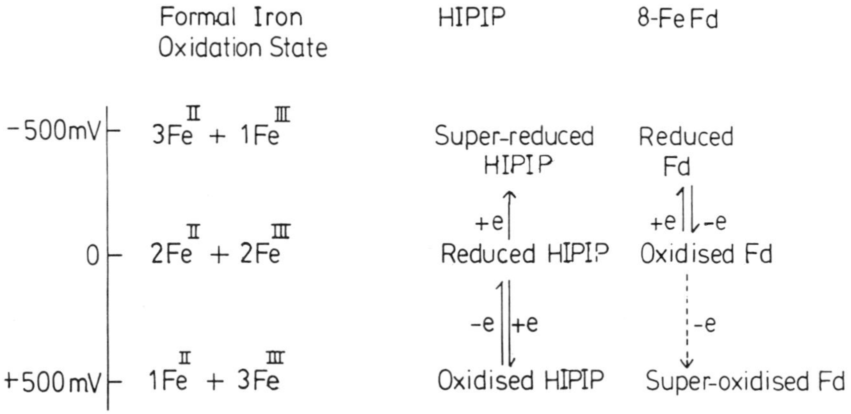

Fig. 7.7 – The three-state hypothesis for the redox potentials of electron transfer in the 4Fe–4S active centres of ferredoxins (Fd) and HIPIP. The redox potentials vary over a 1-V range.

Synthetic Active Site Analogues ('Biomimicry')

Examination of the structures of Rd, Fd and HIPIP proteins shows that these molecules are fundamentally metal complexes with an elaborate ligand structure. As a result Holm and his coworkers considered that analogues of the sites 1–3 might be stable outside a protein environment, and might well be the most thermodynamically stable products in systems containing iron(II) or iron(III), sulphide and/or thiolate anion. This hypothesis has been shown to be the case, and the synthesis of complexes of the general type $[Fe(SR)_4]^-$, $[Fe(SR)_4]^{2-}$, $[Fe_2S_2(SR)_4]^{2-}$, $[Fe_4S_4(SR_4)]^{2-}$ and $[Fe_4S_4(SR)_4]^{3-}$ have been reported. In most of these complexes simple organic thiolate ligands mimic cysteinate binding. Some typical synthetic routes are shown in Fig. 7.8. A number of tetranuclear complexes have also been prepared in which glycylcysteinyl oligopeptides are the terminal ligands; in addition the water soluble complex $[Fe_4S_4(SCH_2CH_2CO_2)_4]^{6-}$ which is isoelectronic with tetranuclear dianions derived from mononegative thiolates has been synthesized. Some typical structures determined by X-ray diffraction are shown in Figs. 7.9 and 7.10.

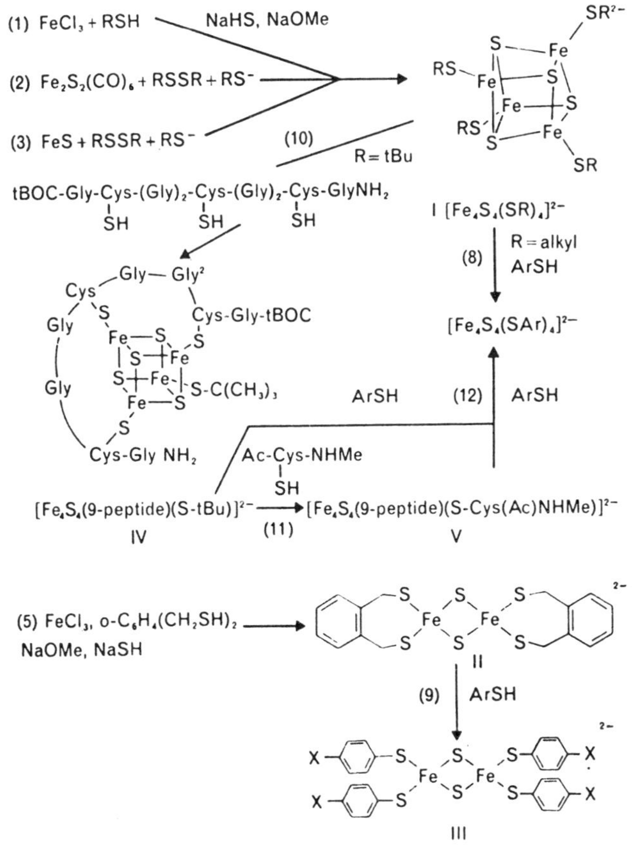

Fig. 7.8 – Synthesis and some reactions of tetranuclear and binuclear Fe–S complexes. The structure of I is drawn so as to conform to the D_{2d} core symmetry found in the R=CH_2Ph and Ph complexes.

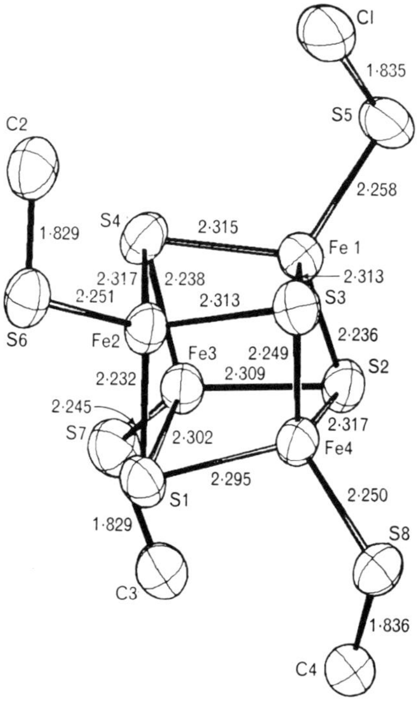

Fig. 7.9 – Structure of $[Fe_4S_4(SCH_2Ph)_4]^{2-}$ (Et_4N^+ salt). Phenyl groups are omitted. The average Fe . . . Fe distance is 2.747 Å. (Reproduced by permission of the National Academy of Sciences).

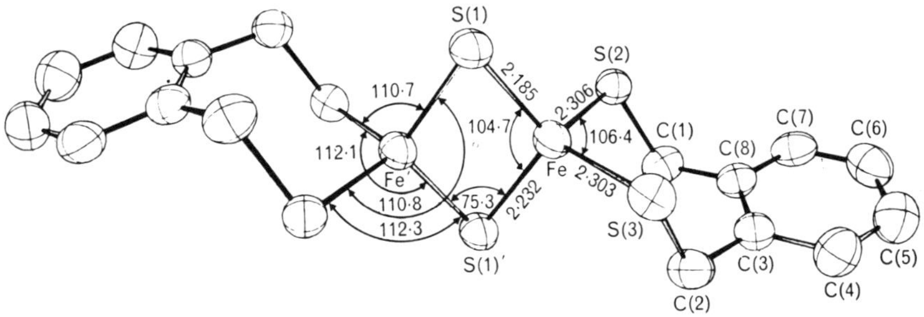

Fig. 7.10 – Structure of bis[(*o*-xylyl-α,α'-dithiolato-μ_2-sulphido ferrate (III)] dianion (II, Et_4N^+ salt). (Reproduced by permission of the National Academy of Sciences.)

Because of the close resemblance of the spectroscopic and magnetic properties of $[Fe_2S_2(S_2\text{-}o\text{-xyl})_2]^{2-}$ with that of the metalloprotein the model complex is considered to be an adequate minimal representation of the structure of the 2Fe-Fd_{ox} site **(7.4)**. A requirement of any synthetic model complex is that it encompasses all known oxidation levels of the corresponding protein site. This point has been established by polarographic techniques using non-aqueous solvents and aqueous-non-aqueous solvent systems. The electron transfer processes shown in Fig. 7.11 have been established. Only the final series is incomplete, lacking $[Fe_4S_4(SR)_4]^0$ with 4Fe(III).

$$\underset{Rd_{red}}{[Fe(SR)_4]^{2-}} \rightleftharpoons \underset{Rd_{ox}}{[Fe(SR)_4]^{-}}$$

$$\underset{(2Fe(II))}{[Fe_2S_2(SR)_4]^{4-}} \rightleftharpoons \underset{(Fe(II)+Fe(III))}{[Fe_2S_2(SR_4)]^{3-}} \rightleftharpoons \underset{(2Fe(III))}{[Fe_2S_2(SR)_4]^{2-}}$$

$$Fd_{super\text{-}red} \qquad Fd_{red} \rightleftharpoons Fd_{ox}$$

$$\underset{(4Fe(II))}{[Fe_4S_4(SR)_4]^{4-}} \rightleftharpoons \underset{(3Fe(II)+Fe(III))}{[Fe_4S_4(SR)_4]^{3-}} \rightleftharpoons \underset{(2Fe(II)+2Fe(III))}{[Fe_4S_4(SR)_4]^{2-}} \rightleftharpoons \underset{(Fe(II)+3Fe(III))}{[Fe_4S_4(SR)_4]^{-}}$$

$$Fd_{red} \rightleftharpoons Fd_{ox} \rightleftharpoons Fd_{super\ ox}$$

$$HIPIP_{super\text{-}red} \qquad HIPIP_{red} \rightleftharpoons HIPIP_{ox}$$

Fig. 7.11.

CORE EXTRUSION REACTIONS

The occurrence of rapid thiolate substitution reactions (due to labile tetrahedral iron centres) offers the possibility of protein reconstitution from the apoprotein and preformed iron-sulphur cores, and removal of intact iron-sulphur cores from proteins. These reactions can be represented as:

$$\text{haloprotein} + RSH \longrightarrow \begin{matrix} [Fe_2S_2(SR)_4]^{2-} \\ \text{and/or} \\ [Fe_4S_4(SR)_4]^{2-} \end{matrix} + \text{apoprotein}$$

A considerable number of these 'core extrusion' reactions have now been carried out, in many cases using benzenethiol(PhSH) as the attacking thiol. The reaction of Rd_{ox} with *o*-xylyl-α, α'-dithiol removes a single iron atom from the protein molecule:

$$Rd_{ox} + 2\text{-}o\text{-xyl(SH)}_2 \longrightarrow [Fe(S_2\text{-}o\text{-xyl})_2]^- + \text{apo-Rd}$$

This latter reaction, can of course, hardly be described as a core extrusion, but illustrates that such reactions can be carried out with all three sites.

Thiolate substitution reactions can also be carried out with synthetic cores. Hence the reactions shown in Fig. 7.12 are readily carried out in DMSO solution.

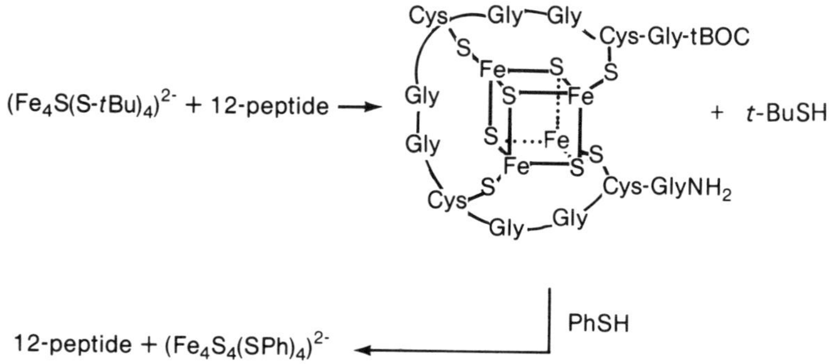

Fig. 7.12.

ELECTRONIC STRUCTURES

The ferredoxins and rubredoxins contain high spin Fe(III) and high spin Fe(II) in the oxidized and reduced forms. The stereochemistry around iron is distorted tetrahedral. The Mössbauer data show that in the oxidized form of the 2-Fe ferredoxins, the two iron atoms are equivalent and are in the high spin $t_{2g}^3\ e_g^2$ configuration. No e.s.r. spectrum is observed in the oxidized form, and there must be an antiferromagnetic interaction between the two $S = 5/2$ Fe(III) atoms. The Mössbauer spectrum of reduced ferredoxin displays two doublets. The most obvious way to allocate the one additional electron is to form a spin–spin coupled Fe(II)–Fe(III) dimer. In the Fe_4S_4 clusters, the configuration for $[Fe_4S_4]^{2-}$ is $Fe_2^{III}\ Fe_2^{II}$. This oxidation state contains an even number of d electrons and therefore an antiferromagnetic interaction between four Fe ions could lead to a diamagnetic state ($S = O$). The $Fe_4S_4^-$ structure (the oxidized state of HIPIP) and $Fe_4S_4^{3-}$(the reduced state of bacterial ferredoxins) is then paramagnetic, $Fe_3^{III}Fe^{II}$ and $Fe^{III}Fe_3^{II}$ respectively.

The Fe–Fe distances are in the range 3.4–3.6 Å, so that the contribution of direct metal-metal orbital overlap to the exchange must occur by overlap of the metal d-orbitals with the bridging sulphur orbitals, i.e. a superexchange pathway.

BIBLIOGRAPHY

Nitrogen Fixation

[1] J. R. Postgate, ed. *The Chemistry and Biochemistry of Nitrogen Fixation,* Plenum Press, 1971.

[2] J. Chatt, J. R. Dilworth and R. L. Richards, 'Recent Advances in the Chemistry of Nitrogen Fixation', *Chem. Rev.,* **78**, 589 (1978).

[3] L. E. Mortenson and R. N. F. Thorneley, 'Structure and Function of Nitrogenase', *Ann. Rev. Biochem.,* **78**, (1979).

[4] R. W. F. Hardy, F. Bottomley and R. C. Burns, eds. *A Treatise on Dinitrogen Fixation, Sections I and II: Inorganic and Physical Chemistry and Biochemistry,* Wiley, New York, 1977.

[5] W. E. Newton and W. H. Orme-Johnson, eds., *Nitrogen Fixation,* Vols. I and II, University Park Press, Baltimore, 1980.

[6] A. H. Gibson and W. E. Newton, eds., 'Current Perspectives in Nitrogen Fixation', Australian Academy of Science, Canberra, 1981.

Dinitrogen Complexes

[7] A. D. Allen and F. Bottomley, 'Inorganic Nitrogen Fixation, Nitrogen Compounds of the Transition Metals', *Acc. Chem. Res.,* **1**, 360 (1968).

Iron Sulphur Proteins

[8] R. H. Holm, 'Metal Clusters in Biology: Quest for a Synthetic Representation of the Catalytic Site of Nitrogenase', *Chem. Soc. Rev.,* 455 (198).

[9] R. H. Holm, 'Iron-sulphur Clusters in Natural and Synthetic Systems', *Endeavour,* **34**, 38 (1975).

[10] R. Mason and J. A. Zubieta, 'Iron-Sulphur Proteins: Structural Chemistry of their Chromophores and Related Systems', *Angew. Chem. Int. Ed.,* **12**, 390 (1973).

[11] R. H. Holm, 'Identification of Active Sites in Iron-Sulphur Proteins', in *Biological Aspects of Inorganic Chemistry,* eds. A. W. Addision, W. R. Cullen, D. Dolphin and B. R. James, John Wiley, New York, 1976.

[12] G. Palmer, in *The Enzymes,* ed. P. D. Boyer, Academic Press, New York, Vol. 12, p. 1, 1975.

[13] 'Iron-Sulphur Proteins', ed. W. Lovenberg, Academic Press, New York, Vols. I, II and III, 1973, 1977.

[14] W. V. Sweeney and J. C. Rabinowitz, 'Proteins Containing 4Fe–4S Clusters. An Overview', *Ann. Rev. Biochem.,* **49** (1980).

CHAPTER 8

Metal Ion Transport and Storage

INTRODUCTION

The earlier chapters discuss the many ways in which metal ions can be exploited to serve the needs of living organisms. To supply these needs, the metals must first be assimilated and then delivered to the sites where they are required. In higher organisms with many types of specialized cells, regulatory mechanisms are necessary to ensure that (a) metal ions are delivered to cells which require them, (b) metals are processed within cells, and conserved and reprocessed at the end of the life span of the molecules or cells containing them and, (c) unrequired metal ions are eliminated, either by excretion or internal storage. Excess of a free metal may produce toxic symptoms by combining non-specifically with proteins or other biomolecules, by distorting the normal metabolism of other metals, and in the case of iron, by causing the formation of damaging free radicals. Detoxification and reserve or transport functions may be combined within the same molecule.

IRON

Iron is an essential trace metal for all living organisms with the exception of lactobacilli. The adult human body normally contains about 3-4 g of iron, of which 70% is in haemoglobin and myoglobin, and intracellular enzymes account for approximately 0.7%. The bulk of the remainder is present as stored iron. Since Fe(III) tends to form insoluble hydroxides at physiological pH (the solubility product of $Fe(OH)_3$ is 10^{-36}), iron(III) is handled by specific transport and storage proteins, transferrin and ferritin.

Transferrin

Transferrin is an iron transport glycoprotein with molecular weight of about 80,000, containing a single polypeptide chain, which is synthesized in the liver. It has two Fe(III)-binding sites, but it is normally maintained at 30% saturation, so that spare sites are available. Each protein molecule contains two identical glycans, ending in sialic acid and conjugated through a GlcNAC–Asn linkage.

Each Fe(III)-binding site also binds an anion, normally carbonate or possibly bicarbonate. The iron(III) and anion are bound and released synergistically.

The presence of two Fe(III) binding sites on a single polypeptide chain immediately raises the question whether the two binding sites are alike or different. A variety of experimental evidence suggests that they are closely similar. Although a great deal of experimental work has been carried out, the nature (i.e. ligand groups) of the binding sites remains unidentified. Spectroscopic measurements (especially e.s.r. spectroscopy) have provided the most useful probes of site conformation. At pH 6.0, binding to a single site is found with Fe^{2+}, VO^{2+} and Cr^{3+} and the second site can be labelled at higher pH. By comparing the spectra of Fe(III) and Cr(III)-transferrins with and without vanadyl at the second site, it has been deduced that Fe(III) binds at the site designated A, while Cr(III) binds at site B at pH 6.0. Even at pH 7.5 transferrin is preferentially labelled at site A by iron. A predominantly Fe_A transferrin can be prepared from [Fe(NTA)] and a predominantly Fe_B transferrin from $[Fe(C_2O_4)_3]$. The half life of transferrin iron in the serum (1-2 h) is a small fraction of that of the protein (7–8 days). The mechanism whereby transferrin releases its iron to the cells is under active investigation.

Ferritin

Ferritin acts as a temporary iron store and as a long-term mobilizable reserve. The iron is in micellar form as an iron(III) oxyhydroxide-phosphate complex. A protein shell (apoferritin) encloses the micelles thus rendering their iron(III) soluble in the cell fluid. Ferritin which contains up to about 4500 Fe atoms usually maintains a reserve capacity for iron. Apoferritin can be conveniently prepared using sodium dithionate in sodium acetate buffer at pH 4.6–5.0. The protein and the Fe(II) so obtained, can be separated by chelation of the Fe(II) to α,α-bipyridyl. Reduction of Fe(III) to Fe(II) is relatively slow and may take several hours. Horse spleen apoferritin has been characterized as a molecule of total M = 444,000 containing 24 subunits each of M = 18,500. A schematic drawing of the quaternary structure of horse spleen apoferritin is shown in Fig. 8.1. Iron enters the shell through channels along four-fold axes. In 1955 it was shown that a ferritin-like product could be formed by the addition of $Fe(NH_4)_2(SO_4)_2$ to apoferritin in air.

It was also found that the apoprotein catalyses Fe(II) oxidation,

$$2\,Fe^{II} + O_2 + 3\,H_2O \longrightarrow 2\,Fe^{III}\,OOH + 4\,H^+ + 1/2\,O_2$$

Ferritin iron(III) can be released by reducing agents such as sodium dithionite or thioglycollate at pH 5 or by cysteine or ascorbic acid. Reduced riboflavins can mobilize iron from ferritin rapidly at neutral pH. The isolation of a NADH-FMN oxidoreductase from rat liver provides a possible biological pathway for releasing ferritin iron through the production of $FMNH_2$.

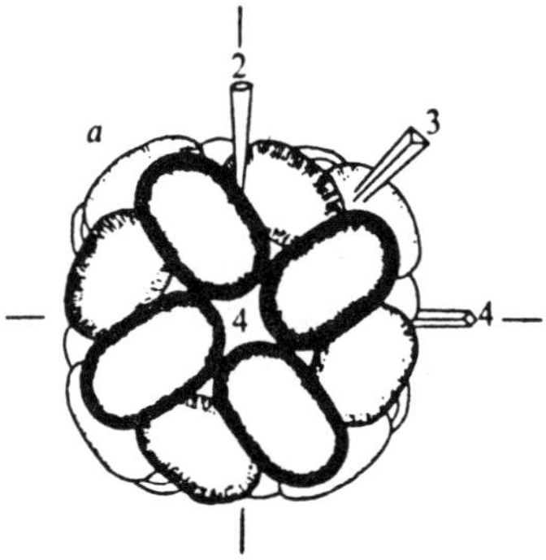

Fig. 8.1 – Schematic drawing of quaternary structure of horse spleen apoferritin, viewed down the four-fold axis. 24 subunits pack together in 432 symmetry to form a shell surrounding a central cavity in which iron may be accumulated. Some of the symmetry axes are shown. Iron enters the shell through channels along four-fold axes. (Reproduced by permission from *Nature*, **271**, 282 (1978).

Siderophores (Siderochromes)

The siderophores (formerly called siderochromes) are low molecular weight compounds synthesized by microbes which are involved in their cellular iron transport. The siderophores are chelating ligands which form extremely stable octahedral complexes with high spin iron(III). Two important classes of these compounds, the ferrichromes and the ferrioxamines, are trihydroxamic acids which (except for the ligands containing charged substituents) form neutral complexes with iron(III). All of the complexes are kinetically labile. The hydroxamic acids have a pK_a of the order of 9 and react with iron(III) to give very

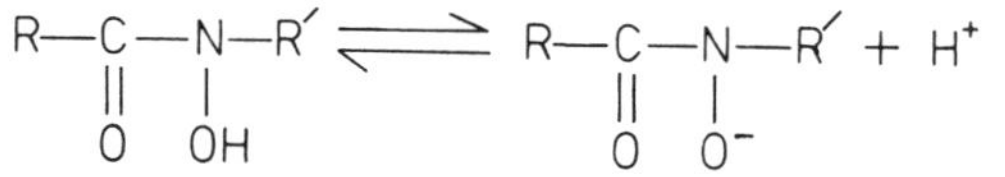

stable five-membered ring complexes (**8.1**). The formation constants for even simple complexes of monohydroxamic acids with Fe(III) are large and quite specific for Fe(III). Thus for acetohydroxamic acid (**8.1**), R = Me, R′ = H), $\beta_3 = 2 \times 10^{28}$, while for the bis complex with iron(II), β_2 is only 3×10^8. The great difference between the chelating abilities of hydroxamic acid for Fe(III) and Fe(II) is probably their most important property for iron transport, since reduction of the Fe(III) complex within the cell provides a means of releasing the complexed iron and releasing the ligands to transport more Fe(III).

The structure of the ferrichromes is shown in Fig. 8.2. The basic structural feature is a cyclic hexapeptide with the three hydroxamic acid groups produced

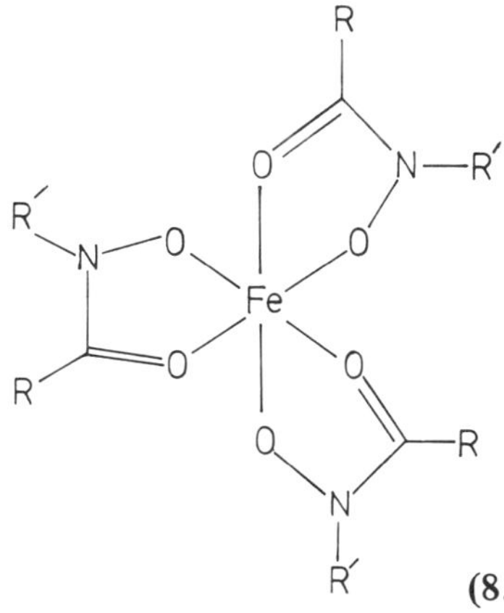

(8.1)

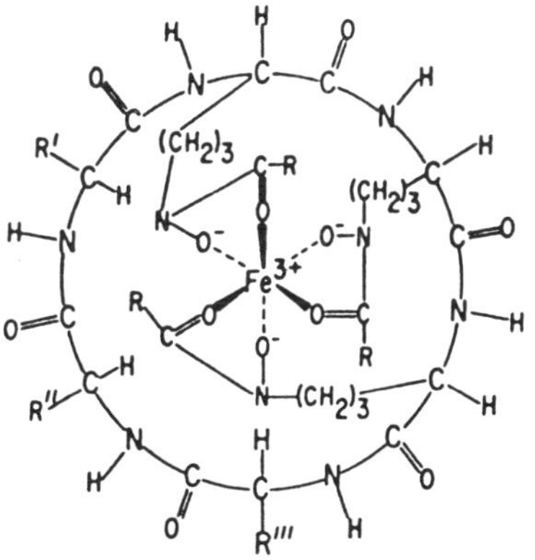

Siderochrome	R′	R″	R‴	R
Ferrichrome	H	H	H	CH_3
Ferrichrysin	CH_2OH	CH_2OH	"	"
Ferricrocin	H	"	"	"
Ferrichrome C	"	CH_3	"	"
Ferrichrome A	CH_2OH	CH_2OH	"	$-CH=C(CH_3)-CH_2CO_2H$ (*trans*)
Ferrirhodin	"	"	"	$-CH=C(CH_3)-CH_2CH_2OH$ (*cis*)
Ferrirubin	"	"	"	$-CH=C(CH_3)-CH_2CH_2OH$ (*trans*)
Albomycin δ_1	$-CH_2OSO_2-$ 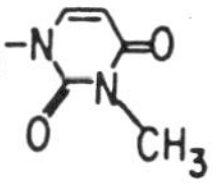	"	CH_2OH	CH_3

Fig. 8.2 – Structure of ferrichromes. The basic structural feature is a cyclic hexapeptide with the three hydroxamic acid linkages provided by a tripeptide of δN-acyl-δN-hydroxyl-S-ornithine. The Λ-cis coordination isomer is shown in each case. From Bioinorganic Chemistry II. Advances in Chemistry Series 162, ed. K. N. Raymond, American Chemical Society, Washington, 1977. Reprinted with permission of the American Chemical Society.

by a tripeptide of δN-acyl-δN-hydroxyl-S-ornithine. Most of the naturally occurring hydroxamic acids have three hydroxamic acid groups per molecule. The three hydroxamate groups are linked either as *pendant* groups from a cyclic peptide (ferrichromes) or as *part* of a linear or cyclic chain (ferrioxamines, Fig. 8.3).

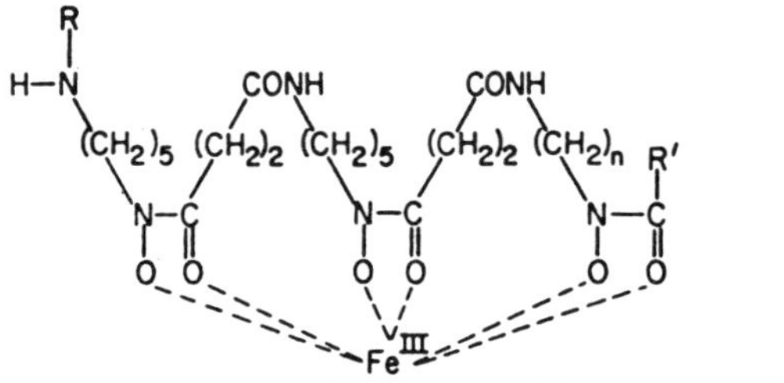

	R	n	R′
Ferrioxamine B	H	5	CH_3-
Ferrioxamine D_1	$CH_3C(O)-$	5	CH_3-
Ferrioxamine G	H	5	$HO_2C(CH_2)_2-$
Ferrioxamine A_1	H	4	$HO_2C(CH_2)_2-$
Ferrimycin A_1	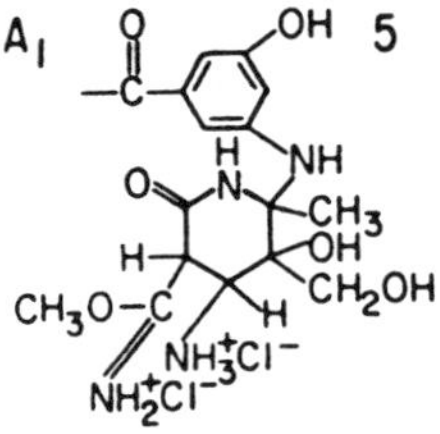	5	CH_3-

Fig. 8.3 – Structure of the linear ferrioxamines. The basic structural feature of the ferrioxamines is repeating units of 1-amino-5-hydroxyaminopentane and succinic acid. Ferrioxamine E is cyclic with $n = 5$ and an amide linkage such that there are no R or R′ substituents, but just a C–N bond. From Bioinorganic Chemistry II. Advances in Chemistry Series 162, ed. K. N. Raymond, American Chemical Society, Washington, 1977. Reprinted with permission of the American Chemical Society.

An iron(III) complex of, for example, ferrioxamine B has five geometrical isomers which are enantiomeric, Fig. 8.4. The Δ-*C-cis, cis* isomer is the mirror image of the Λ-*C-cis, cis* isomer. As the iron(III) complexes are kinetically labile it has not been possible to study the geometrical and optical isomerism of the iron(III) complexes. Raymond and his collaborators have prepared the inert chromium(III) complexes of a number of these ligands and were able to separate the *cis* isomer from one or more of the four possible *trans* isomers. The *cis*

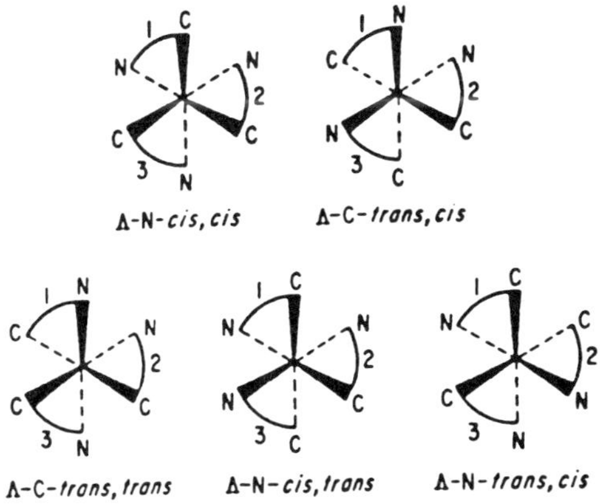

Fig. 8.4 – The five enantiomeric geometrical isomers of ferrioxamine B. The oxygen donor atoms of each hydroxamate group have been omitted for clarity. The Λ optical isomer is shown in each case. (*JACS*, **97**, 294, 1975). From Bioinorganic Chemistry II. Advances in Chemistry Series 162, ed. K. N. Raymond, American Chemical Society, Washington, 1977. Reprinted with permission of the American Chemical Society.

isomer of desferriferrioxamine B has λ_{max} 419 nm (ϵ = 68) and λ_{max} 583 nm (ϵ = 71) while the *trans* isomers have λ_{max} 411 nm (ϵ = 51) and 589 nm (ϵ = 51) and 589 (ϵ = 72). Both the *cis* and *trans* isomers of desferriferrioxamine B-chromium (III) isomerize with half-lives of several days in solution at room temperature.

COPPER AND ZINC

Copper and zinc are both essential to most forms of life. In humans, zinc is quantitatively more important, as there is 1-2 g of the element in the adult human body compared with 0.1 g of copper. As a result of similarities in their chemistry, Zn and Cu share some common ligands, and Zn and Cd may antagonize the absorption of Cu by competing for the same binding sites.

Ceruloplasmin and Serum Albumin

Some 96% of the copper circulating in the blood of normal mammals is present as a blue protein called ceruloplasmin. The protein is synthesized in the liver and contains 6 (or 7) Cu atoms per molecule (M = 132,000). The available evidence suggests that ceruloplasmin acts as a transport protein and that its copper is incorporated into cytochrome oxidase and other enzymes. Some copper is also transported by albumin and this protein is the principal carrier in portal blood between the intestine and the liver. Amino acid-Cu complexes also play a role in Cu transport. Albumin has one specific site for copper, and the copper bound to albumin is in equilibrium with unbound copper or with copper bound by other ligands. It has been suggested that the copper bound to albumin (which is exchangeable) is the immediate transport form of copper.

The albumin molecule consists of a single polypeptide chain of approximately 600 amino-acid residues, with some 17 disulphide linkages. The copper binding site of albumin is considered to involve the α-amino nitrogen, imidazole nitrogen from the histidine in position 3 and the two intervening peptide nitrogens (Fig. 8.5). The structure of ceruloplasmin has been less intensly studied than transferrin. Only about 44% of the copper is paramagnetic. This copper

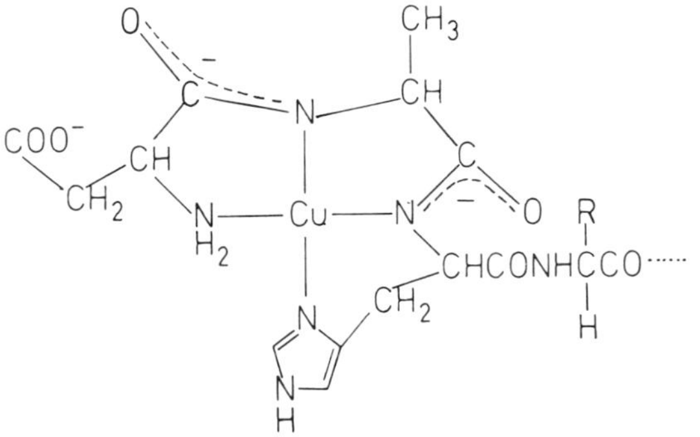

Fig. 8.5 – Structure of the copper-binding site of human albumin.

probably consists of two Type I Cu(II) centres, and one Type II. Type III Cu(II) is diamagnetic with an absorption band at 330 nm. As Type III Cu(II) is thought to consist of two antiferromagnetically coupled Cu(II) ions, there is something of a problem with ceruloplasmin as there appears to be three Cu atoms in this class. The 330 nm absorption of Type III Cu may be due to an R_2S (σ) → Cu^{II} charge-transfer transition of a square-planar or quasi-square pyramidal Cu(II) system bridged to a similar unit. An alternative model for Type III copper is shown in Fig. 8.6, in which there is an N_2S_2 donor system on each copper. Each copper has a distorted tetrahedral stereochemistry, the copper centres being 5–6 Å apart.

METALLOTHIONEINS AND RELATED PROTEINS

A group of low molecular weight proteins (M = 6000–10,000) have been isolated from liver and kidney and other tissues, which have a high affinity for a number of metals especially zinc, cadmium and copper. The proteins (thioneins) have an unusual distribution of amino acids with a high content of cysteine (but not cystine) and little or no aromatic residues resulting in the lack of the

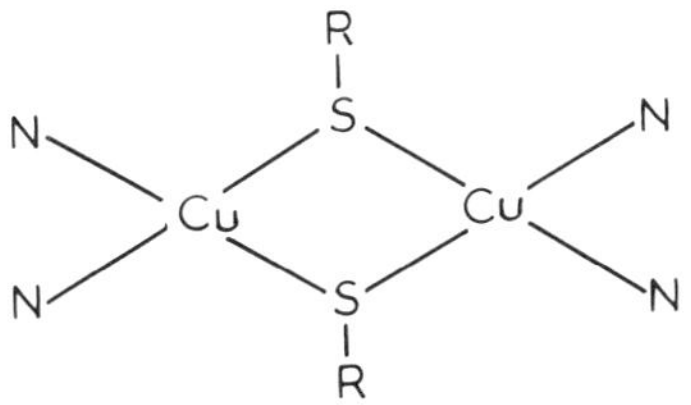

Fig. 8.6 – Possible model for Type III copper(II).

usual protein absorption at 280 nm. The finding that Cd and other non-essential metals, including mercury, bismuth, gold and silver, will not only bind to these proteins, but will induce their synthesis in a number of organs (Table 8.1), raises the question of their function, and suggests that for these metals it is primarily one of detoxification. Zinc and copper thioneins may also be important in the storage and distribution of these essential metals.

Table 8.1 – Increases in thionein-like proteins in various organs in response to various metals.

Metal	Liver	Kidney	Spleen
Cadmium	+	+	+
Zinc	+	—	—
Mercury	—	+	—
Copper	+	+	—
Bismuth	?	+	—
Gold	?	+	—
Silver	+	+	?
Lead	—	—	—

It has been suggested that the term metallothionein is best reserved for those proteins which have a molecular weight of *ca* 6000, and which contain 33 residues per cent of cysteine and which bind 6-7 gram atoms of the metal ion. The primary structure of five thioneins has been reported, their structures are similar and show a striking distribution of cysteine with seven Cys-AA-Cys sequences (AA = one amino acid residue). These portions of the molecule may correspond to the seven, probably separate metal-binding sites. All twenty of the cysteinyl side chains are involved in the metal binding and all the proteins contain 61 amino acids with N-terminal *N*-acetyl methionine.

IRON AND PLANT GROWTH

The amount of iron(III) available in aqueous solutions of most calcareous soils is insufficient for plant growth. Above pH 4 the concentration of iron(III) in solution decreases roughly a thousand-fold for each unit increase in pH. At pH 9, the maximum concentration of iron (both iron(II) and iron(III)) drops below 10^{-20} mol dm^{-3} in a solution in equilibrium with atmospheric oxygen. Plants show varying abilities to respond to iron deficiency (iron stress), and both iron-deficient (chlorotic) and iron-sufficient (green) plants of the same species can grow side by side. The chemical reactions induced by iron stress make iron available to the plant, and plants are classified as iron efficient if they respond to iron stress and iron inefficient of they do not. Iron-efficient plants respond by causing the roots to release H^+ and reducing agents to reduce Fe(III) and to accumulate citrate, thus making iron available to the plant.

METAL TOLERANT PLANTS AND PLANT INDICATORS

As we have seen with iron, metal ions are particularly important for healthy plant life. Excesses or deficiencies of metal ions have effects on plant growth and morphology which are well documented. Excessive concentrations of metals in soils may be caused by the presence of undisturbed ore bodies near the surface (geochemical anomalies), or may be the result of mining. High concentrations can also result from agricultural sources and refuse disposal.

Plants which are diagnostic of particular environmental conditions are known as indicator plants. A variety of plants act as mineral indicators, thus *Lychnis alpina* was used to locate copper in medieval Scandinavia. *Campanula rotundiflora* is associated with smelting sites and lead mineralization. Recently, indicator plants have been used to discover copper in Zambia (*Becium homblei*), and uranium in Colorado using *Astragalus* species. *Astragalus* takes up selenium, which acts as a 'pathfinder' element for uranium.

Plants that grow in soils with metal concentrations which are normally toxic are said to be metal tolerant. Metal tolerant species have been used in attempts to reclaim and recolonize metal-contaminated wastelands. Possible mechanisms for metal tolerance in plants are summarized in Table 8.2. Calcium is an extremely important element in the consideration of the toxic soils. Addition of calcium to metaliferous soils decreases the acidity and reduces the availability of metals such as boron, iron, zinc, nickel, cobalt and manganese.

Plant cells are encased in a wall consisting of cellulose and other polysaccharides. As the plant cell ages, the wall becomes impregnated with other substances, notably lignin. The walls are cemented together by a substance mainly composed of pectates in which Ca^{2+} acts as a structure former. Pectic acid, poly(α-galacturonic acid) is made up from β(1–4) linked α-galacturonic acid **(8.2)**.

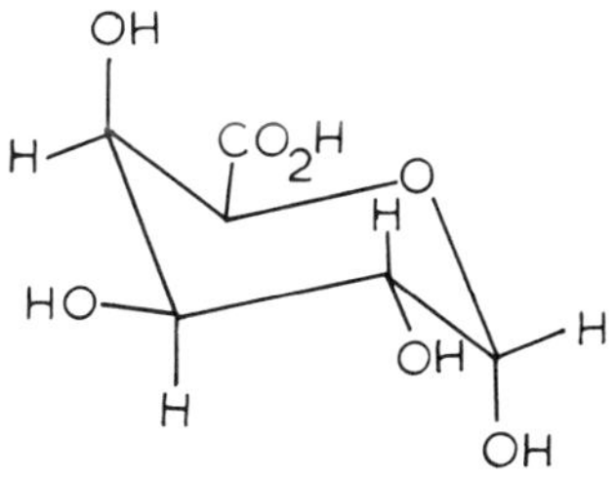

(8.2)

Table 8.2 – Possible mechanisms of metal tolerance in plants.[a]

A. External mechanisms
- (1) Metal is not available to plant root
  - (a) metal is present in water insoluble form
  - (b) metal is present in soluble, but chelated form not available to plant root
  - (c) concentration of freely diffusable metal ions is small

B. Internal mechanisms
- (1) Metal is available to plant root but is not taken up
  - (a) alteration of cell wall membrane of roots giving decreased permeability to toxic metal ion
  - (b) alteration to surface enzymes of roots
  - (c) excretion of substance by root, rendering toxic metal insoluble or unavailable
- (2) Metal is taken up but rendered harmless to metabolism within the plant
  - (a) metal bound in insoluble precipitate or complex
  - (b) metal bound in soluble, innocuous, complex of high thermodynamic stability or low kinetic lability
  - (c) metal removed by spatial separation, e.g. in cell walls or vacuole
- (3) Metal is taken but excreted
  - (a) by loss of collecting organ, e.g. shedding of leaf
  - (b) by guttation
  - (c) by leaching of soluble metal by rain
- (4) Metal ion is taken up but metabolism is altered to accommodate increased concentration of metal ions
  - (a) increase of enzymes inhibited by metal
  - (b) inhibited enzyme systems by-passed
  - (c) alternative metabolic pathway not requiring products of inhibited enzyme system
  - (d) metal required for metabolism

[a]From reference [9].

Ions and water are taken up by the plant via the root. In order to reach the xylem the water and dissolved salts must pass through a membrane (plasmalemma) which probably provides the control for the entry of water and dissolved salts into the plant. Early experiments using coloured salts showed that their diffusion into the root stopped at the endodermis. Electron microscopy has established that in barley roots fed with $UO_2(OOCCH_3)_2$ uranium reaches the casperian strip but not beyond. Metal complexes from plants have been identified in a number of cases. Recently iron has been found as the citrate, nickel(II) in malic and malonic complexes, chromium(III) as the oxalato complex and zinc(II) as the galacturonate. Evidence for the occurrence of a copper–proline complex from the roots of *Armeria maritima* has also been presented. Complexes of this type may play a role in detoxification mechanisms for the various metal ions.

Mugineic acid (**8.3**), an amino acid isolated as an exudate from the roots of water cultured barley, is the first compound shown to play a role in the uptake and transport of iron in higher plants. The ligand is capable of solubilizing $Fe(OH)_3$ within the pH range 4-9. The iron solubilizing action is inhibited by other divalent metals $Cu > Co \geqslant Zn > Mn$ due to competitive equilibria.

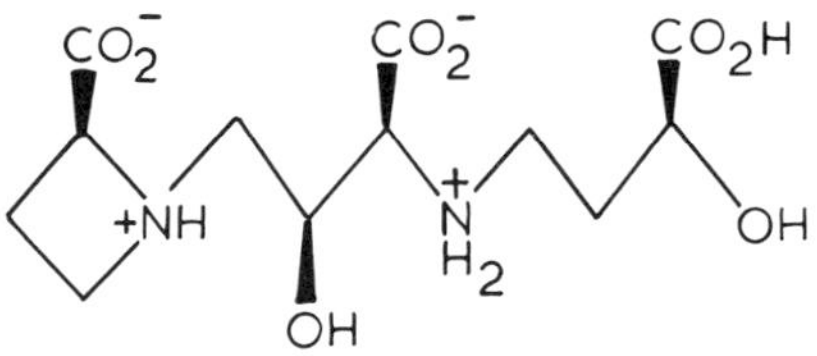

(8.3)

BIBLIOGRAPHY

Ferritin

[1] R. R. Crichton *et al.*, in *Proteins of Iron Metabolism*, ed. E. B. Brown, P. Aisen, J. Feilding and R. R. Crichton, Grune and Stratton, New York, 1977.

[2] P. M. Harrison, *Semin. Haematol.*, **14**, 55 (1977).

[3] R. R. Crichton, *Angew. Chem. Int. Ed.*, **12**, 57 (1973), 'Structure and Function of Ferritin'.

Siderophores

[4] J. B. Neilands, ed. *Microbial Iron Metabolism,* Academic Press, New York, 1974

[5] T. Emery, *Adv. Enzymol.,* **33**, 135 (1971).

[6] J. B. Neilands, *Struct. Bonding,* **1**, 59 (1966).

Metallothioneins

[7] Y. Kojima, and J. H. R. Kagi, *Trends Biochem.,* **3**, 90 (1978).

[8] M. G. Cherian and R. A. Goyer, *Ann. Clin. Lab. Sci.,* **8**, 91 (1978).

Metal Tolerant Plants

[9] M. E. Farago, 'Metal Tolerant Plants', *Coord. Chem. Rev.,* **36**, 155 (1981).

Ceruloplasmin

[10] S. H. Lawrie and E. S. Mohammed, 'Ceruloplasmin: The Enigmatic Copper Protein', *Coord. Chem. Rev.,* **33**, 279 (1980).

Transferrins

[11] P. Aisen and I. Listowsky, 'Iron Transport and Storage Proteins', *Ann. Rev. Biochem.,* **49**, (1980).

Mugineic Acid

[12] K. Nomoto *et al., J. Chem. Soc., Chem. Commun.,* 338 (1981).

CHAPTER 9

Metals and Non-Metals in Biology and Medicine

INTRODUCTION

Of the 25 elements which are currently thought to be essential to life, ten can be classified as trace metal ions: Fe, Cu, Mn, Zn, Co, Mo, Cr, Sn, V and Ni, and four as bulk metals ions: Na, K, Mg and Ca. In addition there is some tentative evidence that Cd and Pb may be required at very low levels. The non-metallic elements are H, B, C, N, O, F, Si, P, S, Cl, Se and I. There is also evidence that Sn, As and Br may possibly be essential trace elements.

In the following sections an outline of the chemistry and biological effects of some of the essential and polluting elements is given. In recent years, a number of metal complexes and ligands have been shown to be chemically useful in a variety of areas, e.g. as antitumour agents, antiviral agents and in the treatment of illness. Various aspects of the chemical application of these compounds will also be described.

METAL POLLUTION

There are a few metals such as cadmium and mercury which have no known physiological function, but many metals are essential nutrients, although they are only required in trace amounts. For example, copper is an essential trace element for all living systems, but if its concentration is raised, it provides one of the most common fungicides and molluscicides. The way in which pollutant metals gain access to the organism and affect the metabolic events within the cell is currently the subject of much study. The picture which is emerging is one in which cells manipulate the transfer of metals across membranes and regulate their enzymic involvement via a series of increasingly specific ligands. It is the disruption of these protein-binding systems by excessive levels of metals which presumably causes the deleterious effects seen in the metabolism of metal polluted animals. Life, however, has evolved in continuous association with metal ions in the environment, and it would be expected that detoxification mechanisms would have been developed that are capable of removing interfering metals.

The cellular detoxification systems in a number of invetebrates have been studied. Two general systems appear to operate. One involves the formation of highly insoluble inorganic granules within vesicules of specific cells. For example, in the garden snail (*Helix aspersa*) the granules contain large quantities of pyrophosphate ($P_2O_7^{4-}$). Pyrophosphate is produced in large quantities by all cells, but it is usually rapidly hydrolysed to orthophosphate (PO_4^{3-}). The use of pyrophosphate provides an excellent system for detoxification, as its salts are normally very much less soluble than the orthophosphates which form similar granules in many other invertebrates. In the snail, special basophil cells in the hepatopancreas accumulate granules of pyrophosphate and if pollutant metals are introduced into the blood of these animals, they are rapidly incorporated into these deposits. The pyrophosphate granules are amorphous and show no clear crystallographic structure.

Inorganic granules of this type occur in virtually every phylum of invertebrate, but they are only effective in removing hard class A metals which like oxygen donors. Soft Class B metals appear to be detoxified by a second system involving the formation of proteins rich in thiol groups. The thionein-like proteins have a strong affinity for metals such as mercury and cadmium and undoubtedly provide protection for the cell against these ions. They may, however, also be involved in the normal metabolism of copper and zinc and play an important role in the supply of these metals for metalloenzyme synthesis.

MAJOR POLLUTANTS

Lead, mercury and cadmium are jointly grouped under this heading. Some 200,000 tons of lead are deposited on the earth annually due to the use of tetraethyl lead and tetramethyl lead in fuels. Alkyl mercury compounds are still present in many lakes at toxic concentrations, however, with the recognition of the hazard, the risk of further alkyl mercury poisoning is decreasing. Cadmium poisoning may increase as the world consumption of the metal increases.

Cadmium

Cadmium accompanies zinc to the extent of about 0.5% in many of its ores, and is obtained as a byproduct of its manufacture. Roughly 5000 tons of cadmium are used annually, mainly in plating, pigments, alkaline batteries and metallurgy. Acute cadmium poisoning (10 mg of cadmium can cause serious symptoms) leads to nausea, salivation, vomiting, diarrhoea and abdominal pain. The toxicity of cadmium taken by mouth is partly offset by the vomiting it frequently induces.

Chronic cadmium poisoning due to long-term, low-level, dosage, is primarily an industrial hazard. A severe outbreak of chronic cadmium poisoning occurred along the Jintsu river of north-west Japan, and was known as Itai Itai disease. The region is near an old disused non-ferrous metal mine and several hundred people died from cadmium poisoning. Cadmium deposition tends to be cumulative in the kidney, with lower concentrations in the liver. Another characteristic of cadmium poisoning is brittleness of the bones.

Cadmium appears to compete with zinc at the active sites of enzymes. The isolation from kidney and liver of a cadmium-containing metalloprotein, metallothionein, suggests that the protein is involved in detoxification processes.

Mercury

Three types of mercury poisoning can be distinguished, mercury vapour, inorganic mercury and alkyl mercury. Soluble inorganic mercury salts are highly toxic. In excess, $HgCl_2$ causes corrosion of the intestinal tract, kidney failure and ultimately death. The use of $Hg(NO_3)_2$ in the felting process resulted in mercury poisoning becoming an occupational hazard of hatters. Symptoms included difficulty in walking, tremors and mental disability. Hence the expression 'mad as a hatter'. Mercury(II) binds to thiol groups, thus almost all proteins can bind mercury to some extent and are potential targets for mercury poisoning. The importance of methyl mercury, CH_3Hg^{II} in pollution of the environment by mercury became apparent in the 1960s following the surprising discovery that a large fraction of the mercury in fish was methyl mercury, even though some of the fish were taken from lakes and rivers in which no methyl mercury was discharged. Subsequent studies of the biological cycle of mercury revealed chemical and microbiological pathways by which CH_3Hg^{II} can be formed from Hg(II). Thus Me-B_{12} reacts with $HgCl_2$

$$\text{Me-B}_{12} + \text{HgCl}_2 \longrightarrow \text{MeHgCl} + \text{B}_{12} + \text{Cl}^-$$

to give methylmercury chloride. Methylmercury reacts with thiol groups to give very stable complexes. Thus log $K = 22$ with mercaptoalbumin and log $K = 15.7$ with cysteine. Quite stable complexes are also formed with nitrogen donors, thus log K for glycine is 7.88.

Lead

Modern consumer use of lead throughout the industrialized world has more than doubled during the last 30 years. North America today produces approximately 1 million tons of lead annually, or about 10 lbs per inhabitant. The battery industry is one of the largest single users of lead, but leaded petrol accounts for more than 20% of the total lead consumed per year. Tetraethyl lead (TEL), tetramethyl lead (TML) and mixed lead alkyls are used as antiknock additives to improve the combustion characteristics of gasoline. The chemical composition of exhaust particles from internal combustion engines appears to be related to particle size. The major lead products emitted in particles of 2–10 μm equivalent diameter are lead bromochloride (PbClBr), and the α- and β- forms of ammonium chloride and lead bromochloride ($NH_4Cl.2PbClBr$; $2NH_4Cl.PbClBr$), together with minor quantities of lead sulphate and the mixed oxide (PbO.

$PbClBr.H_2O$). The particles subsequently lose halogens (a process which appears to be photochemically enhanced), and become smaller and more soluble (especially in the presence of SO_2). Several reports have suggested that lead halogen compounds are then converted to oxides and/or carbonates. The phased reduction in the lead content of petrol in the UK from the 1971 limit of 0.84 g dm^{-3} to 0.46 dm^{-3} today has coincided with an increase in traffic density. Generally in the UK there appears to have been little change in airborne lead levels over the last 5 years. At the kerbside of busy city streets (10,000+ vehicles per day) the concentration of airborne particulate lead lies within the range 2-5 $\mu g/m^{-3}$; city areas not immediately influenced by traffic have mean lead levels less than 1 $\mu g/m^{-3}$.

Diet is the major source of lead in man. Clinically evident lead poisoning most frequently results from the absorption of lead through the gastrointestinal rather than the respiratory tract. Chronic lead poisoning was a major cause of illness throughout the period of the Roman Empire. The principal source of contamination has been the use of lead compounds in the manufacture, storage, transport and cooking of foodstuffs, as well as the use of lead-based agricultural insecticides. The daily intake of lead from drinking water for an adult is usually 15-20 μg where domestic lead plumbing is still in use, and water is soft and acidic; more than one-third of the total lead intake could be from water. In some areas of Scotland where plumbosolvent water is not treated, the daily intake of lead from drinking water, cooking water, etc., may be excessive, particularly for infants. The WHO-recommended limits for lead are 100 μg dm^{-3} for municipal water supplies.

Retention and Distribution of Lead

Continuous exposure to lead results in its gradual accumulation in the body. The absorbed lead is distributed throughout the body via the blood stream, and is subsequently excreted mainly in the urine. Retention of lead by adults is variable (6-33 μg/day). Estimates of total body burden of lead in adults range from 50-400 mg with a mean of less than 200 mg. The highest concentrations of lead are found in the aorta, liver and kidneys. More than 95% of the total body burden in man is stored in the bone as the relatively insoluble triphosphate. No beneficial biological role for lead has yet been demonstrated and there is as yet no substantial evidence to suggest that it is an essential trace element. At the cellular level lead is found to interfere with the respiratory pigments, energy production and membrane function.

MINOR POLLUTANTS

Aluminium

Although aluminium occurs abundantly in nature, it is normally present as chemical entities (e.g. aluminosilicates) which are quite inert to further reactions.

Aluminium has, however, been implicated in the dialysis encephalopathy syndrome, where aluminium gels provide a source of the aluminium cation. Aluminium has also been shown to cause neuron degeneration and has been found in high concentrations in the brains of patients suffering from Alzheimer's disease, a form of senile dementia.

It has been shown that aluminium inhibits yeast or brain hexokinases and that activation can be restored by various ligands, such as phosphate or carboxylate. Because of the preponderance of aluminium cookware, aluminium (as $[Al(OH_2)_6]^{3+}$) is ingested by a large proportion of the population. Those working in the aluminium industry may ingest it in other forms. The subsequent interactions of the aluminium are of concern. Studies of the interaction of $[Al(H_2O)_6]^{3+}$ with the salts of long-chain fatty acids (by nmr techniques) are beginning to appear and our knowledge of these systems should increase in the near future.

Table 9.1 – Some metal dependent conditions

Essential or beneficial element	Disease arising from deficiency	Disease associated with an excess of the element
Calcium	Bone deformities, tetany	Cataracts, gall stones, atherosclerosis
Cobalt	Anaemia	Coronary failure, polycythaemia
Copper	Anaemia, kinky hair syndrome	S.A.K. Wilson's disease
Chromium	Incorrect glucose metabolism	
Iron	Anaemias	Haemochomatosis, siderosis
Lithium	Manic depression	
Magnesium	Convulsions	Anaesthesia
Manganese	Skeletal deformities gonadal dysfunctions	Ataxia
Potassium		Addison's disease
Selenium	Necrosis of liver, white muscle disease	Blind staggers in cattle
Sodium	Addison's disease, stoker's cramps	
Zinc	Dwarfism, hypogonadism	Metal fume fever
Polluting element		
Cadmium	–	Nephritis
Lead	–	Anaemia, encephalitis, neuritis
Mercury	–	Encephalitis, neuritis

ESSENTIAL TRACE METAL IONS WHICH ARE TOXIC IN EXCESS AND METAL DEFICIENCY

Although ions of cobalt, copper, iron, zinc and manganese are essential for mammals, they become hazardous when present in excess. As a result there are metabolic disorders connected with both deficiencies and excess amounts of these metal ions (Table 9.1).

Copper

Wilson's disease leads to an excessive accumulation of copper in the liver, kidney and brain which leads to liver and kidney failure and various neurological abnormalities. Death results if the condition is not recognized and treated. A genetic defect leads to abnormalities in normal copper metabolism.

The treatment of Wilson's disease involves the administration of various chelating agents (Fig. 9.1) which are capable of mobilizing the copper.

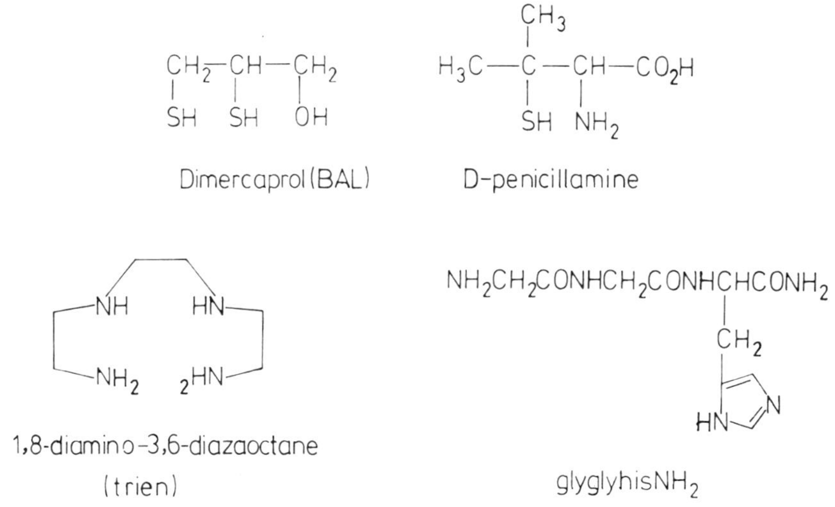

Fig. 9.1 – Chelating agents used in the treatment of Wilson's disease.

Cobalt

Insufficient levels of cobalt in the diet of ruminants gives rise to wasting diseases, known in Australia and New Zealand as bush sickness. Trace amounts of vitamin B_{12} are essential for the synthesis of haemoglobin by mammals.

During the period from 1965 to 1966, cobalt(II) sulphate was added to beer in Quebec (3 mg per gallon) to improve foam stability. Cardiomyopathy in heavy beer drinkers was observed and there were more than 20 deaths. Implantation of cobalt powder has been reported to cause malignant tumours in muscles.

Iron

Acute iron poisoning leads to vomiting, pallor, shock, haematemesis, circulatory collapse and coma. Chronic conditions are also known in which iron is deposited in tissues and organs of the body. These conditions are classified as *haemochromatosis* (iron deposition and hepatic cirrhosis) and *haemosiderosis* (iron deposition, but no cirrhosis).

Manganese

In the form of its salts, manganese is not appreciably toxic when taken orally, but high levels interfere with phosphate retention. Inhalation of manganese oxide dust gives rise to locura manganica which resembles schizophrenia.

Molybdenum

Molybdenum is the only element of the second and third transition series known to be essential for life. Molybdenum is probably absorbed into biological systems as the molybdate anion (MoO_4^{2-}). As with other trace elements which are absorbed anionically, the levels of Mo in organisms are determined primarily by the available Mo in the soil and/or the food and water. Soil molybdenum deficiencies arise from a combination of factors such as good drainage, low pH and/or certain soil structures with a low molybdenum content. Legumes, citrus, broccoli and even grasses and grains planted in such soils show stunted growth and other symptoms which are relieved by the application of a fertilizer containing molybdenum. Plants can incorporate up to several hundred parts per million of Mo into their tissues without showing adverse symptoms. However, inhabitants of the molybdenum-rich Ankavan region of the USSR, who are estimated to consume 10–15 mg/day of Mo, have an unusually high incidence of gout.

NON-METALS

Silicon

Until some 10 years ago, few chemists believed that silicon compounds would have important biological effects. Groups in the USSR and France have subsequently found that many organosilicon compounds have high and specific biological activity. Voronkov and his collaborators in the Soviet Union have shown that the silatranes (**9.1**) are substances with very high and specific biological activity. They exhibit a strong analeptic effect and are highly toxic towards warm-blooded animals. In the USA, M & T Chemicals are now marketing 1-(*p*-chlorophenyl)silatrane as a rodenticide. The compound is rapidly and completely inactivated in poisoned rodents, so that their corpses are not harmful to other animals. On the skin of man, solid 1-aryl silatranes have no toxic effects. The toxicity of these compounds is highly selective. For example, 1-(*p*-chlorophenyl)silatrane is 50 times less toxic towards monkeys than sparrows.

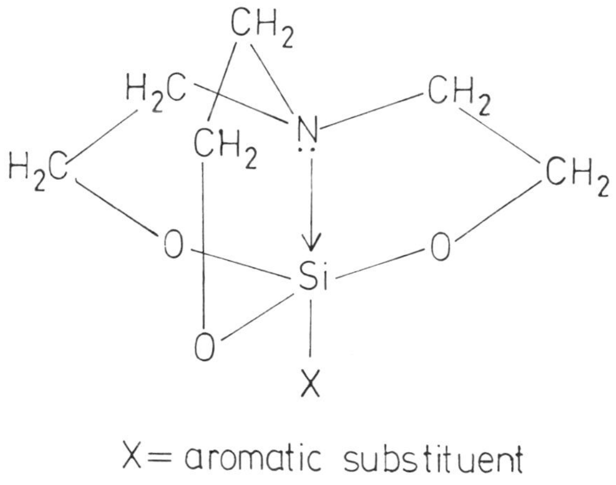

(9.1)

Table 9.2 – New biologically active organosilicon compounds

Type of effect	Class of organosilicon compounds*	Type of effect
Impairment of co-ordination of movement	$R_3SiZNR'R''$ $R_{4-n}Si(OCH_2CH_2NR'_2)_n$ $(R_3SiOCH_2CH_2)_nNH_{3-n}$	Insecticidal activity
	$RSi(OCHR'CH_3)_3N$	Chemosterilizing activity
Reduction of blood pressure	$[R_3SiCH\ N(CH_3)_2CH_2CH_2X]^+Y^-$	
	$R_3SiC{\equiv}C{-}C(OH)R'R''$	Fungistatic activity
Stimulation of breathing	$RSi(OCHR'CH_2)_3N$	
	$[R_3SiCH_2N(CH_3)_2CH_2CH_2X]^+Y^-$	Insect repellant activity
	$[R_3SiOCH_2CH_2N(CH_3)]^+X^-$	
	$R_3SiCH_2CH(COCH_3)CH_2NRR'$	Antibacterial activity
	$R_3SiC{\equiv}C(OH)R'R''$	
Soporific activity	R_3SiOH, R_3SiNCO	Zoocidal activity

*R, R′, R″ = Hydrocarbon or alkoxyl radical; Z = saturated or unsaturated, substituted or unsubstituted three-carbon chain; X, Y = halogen, OH, OCOR′.

Silicon is one of the principal elements in protoorganisms and is important in plants and animals at a low stage of evolutionary development. It participates in particular in the life cycle of many bacteria (silicate bacteria and those living in hot springs). There are some microorganisms (e.g *Proteus mirabilis*) which synthesize compounds of silicon containing Si-O-C, Si-N-C and even Si-C bonds. These bacteria are of interest because of their ability to substitute phosphorus for silicon.

Silicon compounds are contained in the organisms of protozoa (*Foraminiferae*), some algae (*Diatomea*), lichens, cereals (rice), millet, barley, bamboo, feather grass, reed), and in the needles and leaves of some trees (larch, palm, etc). Volatile organosilicon compounds have been found in swede seeds and tobacco smoke. Specific enzymes have been found in plants which catalyse the conversion of mineral silicon into organosilicon compounds.

Silicon appears to increase the viability and crop capacity of plants relatively rich in silicon ('siliceous plants'), also their stability to fungal infections, drought and radiation.

In the tissues of humans and animals silicon compounds appear to be present in three principle forms:

(a) Water soluble inorganic compounds capable of passing through cell walls, which can be readily eliminated from the organism (orthosilicic acid, and ortho-and oligo-silicate ions). Silicic acid ($Si(OH)_4$) is a weak acid $pK_1 = 9$ and $pK_2 = 12.5$. The solubility in water at 20°C is *ca.* 100 p.p.m. The acid condenses to form progressively the dimer, trimer, etc., and eventually polysilicic acid as hydrated silica. The pH of maximum condensation is around pH 5.5.

(b) Organosilicon compounds and complexes soluble in organic solvents and containing Si-O-C groups (ortho- and oligosilicic esters of carbohydrates, proteins, steroids, choline, lipids and phospholipids).

(c) Insoluble silicon polymers (polysilicic acids, silica, silicates) whose surface is always covered with a chemisorbed layer of organic substances. The first organosilicon compound found in an animal organism was the orthosilicic acid ester of cholesterol isolated from bird feather.

Nitrogen

Nitrogen is a constituent of numerous naturally occurring compounds such as amino acids, proteins and nucleic acids. These complex molecules are built up from smaller organic molecules, but the original source of the element lies in simple inorganic compounds. The 'nitrogen cycle' (Fig. 9.2) represents the transformation of inorganic nitrogen to organic nitrogen together with the reverse degradation process. In this cycle, soil ammonia is converted to nitrate by the action of microorganisms (nitrification). Nitrate is assimilated by plants and built into organic molecules. Higher species feeding on these plants convert

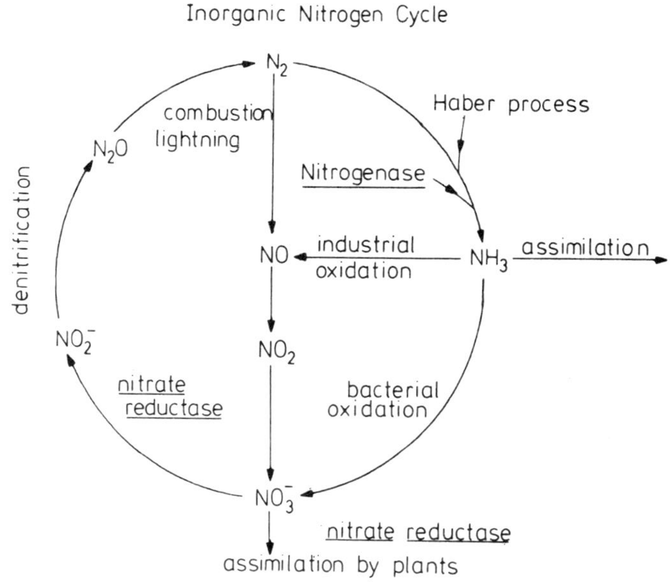

Fig. 9.2 – The nitrogen cycle. The underlined enzymes contain molybdenum.

such molecules into more complex compounds. On death, these compounds are broken down, leading to the reformation of ammonia. Denitrification is the process whereby nitrates are reduced by microorganisms, ultimately N_2 being produced. As we have seen in Chapter 7, atmospheric dinitrogen is fixed (i.e. converted to ammonia) by various bacteria.

A number of intermediates must lie between NH_3 and NO_3^- in both nitrification and denitrification pathways. Hydroxylamine and nitrite (nitrogen in formal oxidation states −I and −III respectively) are now well established as intermediates, but the nature of the species with nitrogen in the oxidation state (+I) is less certain.

Phosphorus

Phosphorus plays a vital role in all life forms, and phosphate esters and diesters are the principal mode in which it performs its essential functions. There are few derivatives with phosphorus–carbon and phosphorus–nitrogen bonds and these are noteworthy for their rarity. The first phosphorus–carbon compound identified was 2-aminoethylphosphoric acid, $NH_2CH_2CH_2P(O)(OH)_2$, which was isolated from certain protozoa, and this remains the predominant example of a naturally occurring P–C compound. Phosphocreatine is an essential P–N compound and this bond may well be the reason for its usefulness as a stored form of phosphate. This compound (9.2) is held in reserve as a reagent for the 'emergency" regeneration of the one biophosphorus compound on which life depends, adenosine triphosphate (ATP).

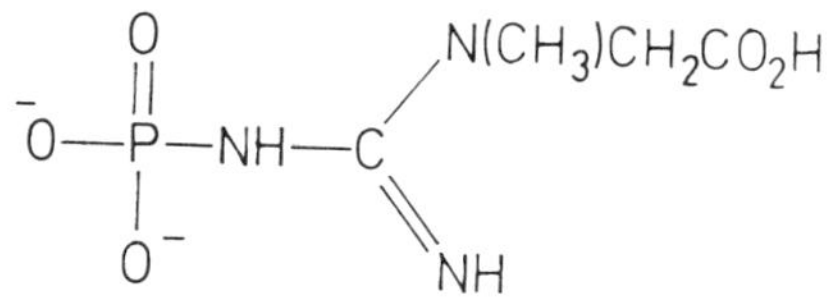

(9.2)

Life forms at the beginning of the food chain absorb phosphorus as inorganic phosphate from soil or water. The inorganic phosphate has to be converted to esters, a reaction which cannot be performed directly as preliminary activation is required. Activation is generally achieved by formation of an anhydride of the type found in ATP.

Phosphate esters have quite large standard free energies of hydrolysis $\Delta G^{0'}$ (pH 7.0, 37 °C and in the presence of excess Mg(II) ions) as shown in Table 9.3. The standard free energies of hydrolysis are useful in determining the direction of a reaction, and in an equilibrium situation giving a value of the equilibrium constant. Thus for the equilibrium:

$$\text{Ribulose-5-phosphate} + \text{ATP} \overset{K}{\rightleftharpoons} \text{ribulose-1,5-diphosphate} + \text{ADP}$$

the overall $\Delta G^{0'}$ is -17.1 kJ mol^{-1} and since $\Delta G^{0'} = -RT \ln K$, $K = 10^3$. ($\Delta G^{0'}$ for ribulose-1,5-diphosphate to ribulose-5-phosphate is $-$ 13.4 kJ mol^{-1}, while for $ATP^{4-} \rightarrow ADP^{3-} + PO_4^{3-}$ $\Delta G^{0'}$ is -30.5 kJ mol^{-1}; the overall $\Delta G^{0'} = -30.5-(-13.4) = -17.1$ kJ mol^{-1}).

One interesting feature of the chemistry of phosphate esters, anhydrides and acyl phosphates is that their hydrolytic reactions are all subject to catalysis

Table 9.3 – Standard free energies of hydrolysis of biophosphorus compounds ($\Delta G^{0'}$)†

Compound	$\Delta G^{0'}$ (kJ mol^{-1})
Phosphoenolpyruvate	−62.0
Phosphocreatine	−43.1
Acetyl phosphate	−42.3
Diphosphate	−33.5
Acetyl CoA	−31.4
ATP^{4-} ($\rightarrow ADP^{3-} + PO_4^{3-}$)	−30.5
ATP^{4-} ($\rightarrow AMP^{2-} + P_2O_7^{4-}$)	−30.5
Glucose-1-phosphate	−21
Glucose-6-phosphate	−14

†$\Delta G^{0'}$ relates to pH 7.0 and 37 °C.

by metal ions such as Mg(II). Due to the large number of negative charges on phosphate ester species, attack by nucleophiles is not favoured on electrostatic grounds. Hard metal ions such as Mg(II) form complexes with a lower overall negative charge and thus provide electrostatic facilitation of hydrolysis.

CANCER

Cancer is a disease characterized by uncontrolled multiplication and spread within the organism of apparently abnormal forms of the organisms own cells. Our bodies produce about 10^{11} cells per day (about 500 g of new tissue), between 10^4 and 10^6 of which are imperfect, but these are generally destroyed by our immune systems.

Some metals are known to be carcinogenic (Be, Y, Cd and some Ni salts). The following metals and non-metals all accumulate in our bodies during ageing: Al, As, Ba, Be, Cd, Cr, Au, Ni, Pb, Se, Si, Ag, Sr, Sn, Ti and V, and their effects are unknown. It has been observed that the distribution of some cancers can be correlated with topsoil trace-metal content. Many organic carcinogens are also excellent ligands (Fig. 9.3), and the importance of coordination compounds in cancer initiation is the subject of considerable debate [17].

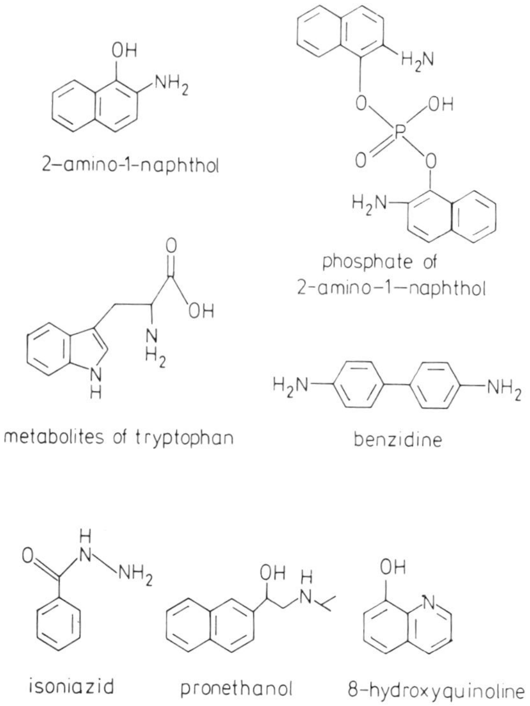

Fig. 9.3 – Some known and suspected carcinogens.

INORGANIC DRUGS

As little as 100 years ago, most drugs were crude preparations obtained from plant, animal or mineral sources, few of which were specific or particularly effective in the relief of symptoms. Even in 1930 the range of drugs available was quite limited. The discovery of the sulphonamides sparked off major developments in the drug industry, and today more than 30,000 drugs are available, some 8000 of which are commonly prescribed.

Although inorganic compounds were initially widely used in therapy (e.g mercury for syphilis, Mg^{2+} salts for intestinal treatments, Fe^{2+} salts for anaemia), the advances made in organic chemistry overshadowed the early interest in inorganic derivatives.

The discovery in the 1950s that some metal complexes were antiviral agents stimulated renewed interest in inorganic derivatives and their role in chemotherapy. UK drugs sales within the last decade have followed the general order, antibiotics > tranquillizers and antidepressants > corticosteroids > cardiovasculars > antiarthritics > cold preparations > analgesics ≡ diuretics ≡ sedatives ≡ oral contraceptives > antacids ≡ vitamins > antianaemics ≡ antibacterials ≡ dermatology preparations.

The pattern reflects the incidence of disease, hence antibiotics have a wider market than antibacterials. Some recent developments in the inorganic area include the use of lithium salts in the treatment of manic depression, the use of Pt(II) compounds in cancer therapy, and gold compounds in the treatment of rheumatoid arthritis.

Lithium Salts

Lithium carbonate is used in the treatment of manic-depressive psychoses. Manic depression involves recurrent periods of *mania* (elevated mood and increased aggression) and depression. Lithium carbonate is used by one in 2000 of the population of the United Kingdom (one in 1000 in the city of Edinburgh), and this is probably about one-half of those likely to benefit from its prophylactic effects. Lithium carbonate tablets are taken (usually 1-2 g per day) to maintain a plasma lithium concentration of 0.5-1.0 mmol dm^{-3}. Symptoms of toxicity are observed at about 2 mmol dm^{-3} leading to course tremor of the hand, polyuria, vomiting, diarrhoea, sluggishness, vertigo and slurred speech. Definitive evidence as to the mode of action of lithium is currently unavailable, but much literature on the topic exists [11–13].

Gold

The biological use of gold can be traced back as far as the Chinese in 2500 BC. However, the observation of Koch in 1890 that gold cyanide inhibited the growth of tuberculosis bacilli represents the beginning of systematic gold pharmacology and of attempts to design gold drugs [10].

The use of gold(I) thiomalate in the treatment of rheumatoid arthritis dates from the early 1960s. Most gold complexes used medically are thiol complexes (Fig. 9.4). Recently, successful studies of the use of some phosphine complexes for gold therapy have been reported; unlike standard mercapto-gold drugs which have to be injected, the phosphine complexes are administered orally.

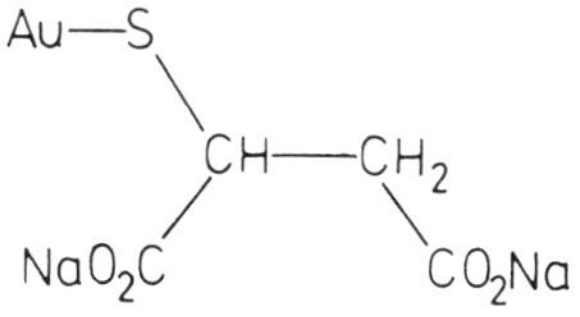

(1)

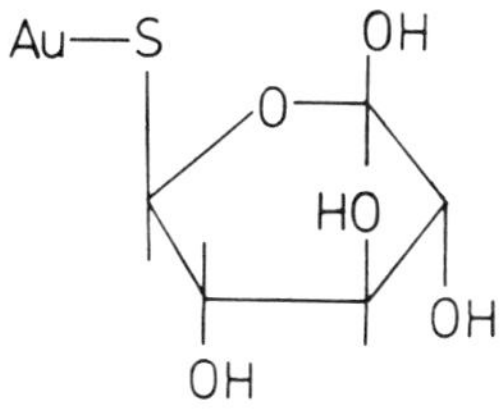

(2)

Et_3—P—AuCl

(3)

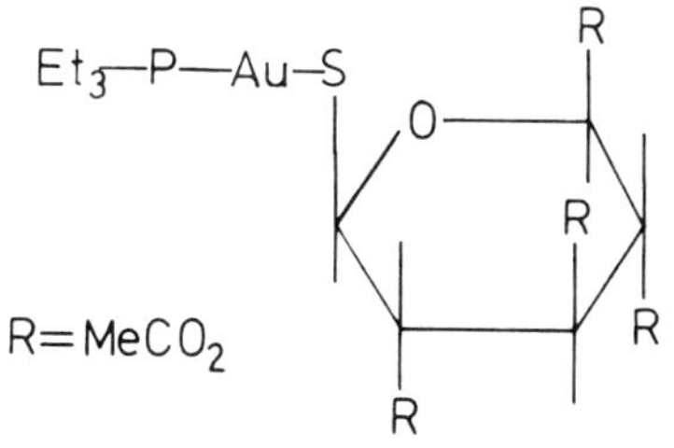

(4)

Fig. 9.4 – Some gold complexes commonly used in the treatment of rheumatoid arthritis. Compounds (1) and (2) are conventional gold drugs which are usually administered by intramuscular injection. Compound (3) is an orally effective form of gold in animal models of inflammation. As a result of further development compound (4) was recently tested as an orally active drug in a clinical trial.

The chemistry of gold *in vivo* is likely to be very different from that of transition metals such as copper and iron which are essential to man. With these latter elements there are carefully controlled transport storage and enzyme functions but there appear to be no similar systems for gold. It seems possible that most of the chemistry *in vivo* is concerned with the reaction of gold species with thiols. Gold(0), gold(I) and gold(III) can occur in biological systems, and both monomeric and polymeric gold species are to be expected. Much remains to be learned of the basic solution chemistry of gold complexes [10].

Platinum

Some complexes of Pt(II) and Pt(IV) are potent antitumour agents, being effective against transplanted, carcinogen-initiated, and virally induced cancers. The complex *cis*-[$Pt(NH_3)_2Cl_2$] ('Cisplatin') can cause complete tumour regression in mice. Furthermore, the treatment also conferred immunity to rechallenge from the same tumour for up to 12 months. Typical examples are shown in Fig. 9.5. The complexes have a *cis* configuration, the *trans* isomers being inactive. In addition the complexes are neutral and are therefore able to cross membranes *in vivo*. The anticancer activity is due to inhibition of DNA synthesis in the cancer cell. Displacement of the chloride ligands by the nitrogen donors of two purines in the DNA chain forms a cross-link. Alkylating agents such as the nitrogen mustards are thought to form similar cross-links, but in this case, between bases on different DNA chains. The distances between the chlorides on the Pt(II) complexes, 3.3 Å and on the nitrogen mustards 8.0 Å, fit perfectly for the formation of intra- and interstrand bridges.

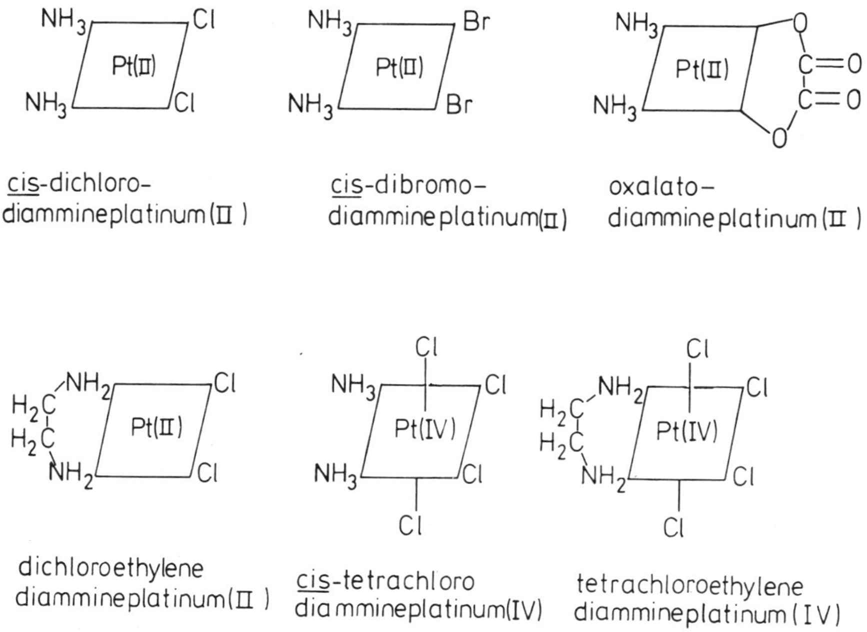

Fig. 9.5 – Structural formulae of some active antitumour complexes of platinum.

Rhodium

Tetrakis(μ-carboxylato)dirhodium(II) **(9.3)** has been shown to function as an antitumour agent by inhibiting DNA synthesis.

Crystallographic studies indicate, that for adenine,N(7) coordination is favoured because of the possibility of intramolecular H-bonding between the

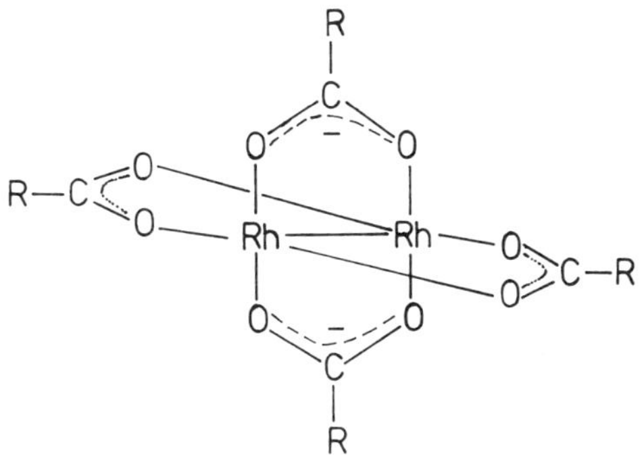

(9.3)

amino nitrogen at C(6) and the carboxylate oxygen as in **(9.4)**. In these complexes the Rh–Rh bond length is 2.412 Å which would allow an intrastrand bridge to form with appropriate DNA bases. *Trans*-$[RhL_4X_2]Y$, where L is a substituted pyridine, X is chloride or bromide and Y = Cl^-, Br^-, NO_3^- or ClO_4^- have high levels of antibacterial activity against Gram-positive organisms and *E. coli.*

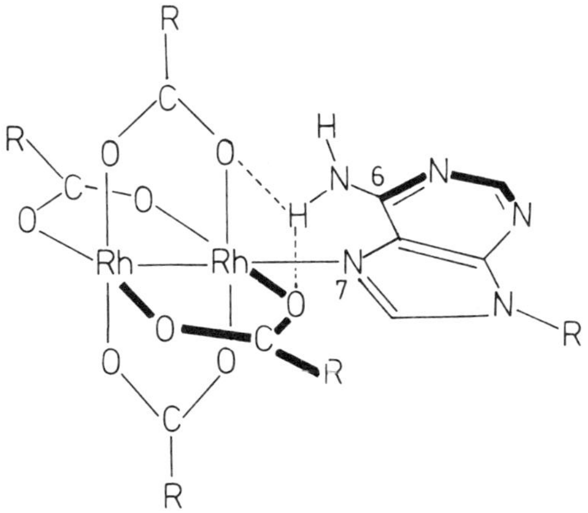

(9.4)

Arsenicals

The discovery of Ehrlich in 1909 of the spirochaeticidal activity of salvarsan (**9.5**) led to the development of a wide range of arsenicals for the treatment of syphilis. These compounds owe their effectiveness to their partial conversion to arsenoxides, followed by reaction of the arsenic with thiol groups of enzymes, particularly pyruvate oxidase and lypoic acid dehydrogenase, leading to inhibition of cellular metabolism. The arsenicals have now been superseded by antibiotics such as penicillin.

Arsenicals are still used in the treatment of other parasitic infections. Derivatives of benzene arsonic acid $C_6H_5AsO(OH)_2$ being commonly used.

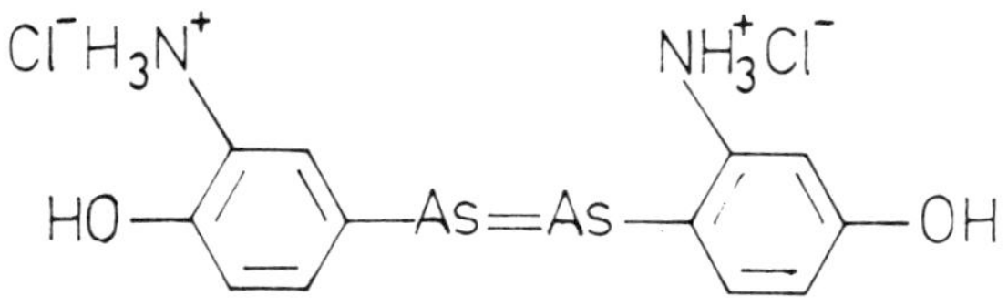

(9.5)

Copper

The earliest fungicides used on plants were sulphur and inorganic copper compounds and the latter still find major agricultural use, particularly copper oxychloride and copper sulphate. These compounds function by forming a sulphate layer in plants which inhibits the growth of any fungal spores that alight on it. The copper(II) complex of 8-hydroxyquinoline is used extensively to rot-proof canvas and has limited application as an agricultural antifungal spray.

BIBLIOGRAPHY

Biology of Metals and Non Metals

Aluminium

[1] A. S. Tracey and T. L. Brown, *J. Am. Chem. Soc.*, **105**, 4901 (1983).

Lead

[2] M. E. Hilburn, 'Environmental Lead in Perspective', *Chem. Soc. Rev.*, **8**, 63 (1979).

Selenium

[3] T. C. Stadtman, 'Selenium Dependent Enzymes', *Ann. Rev. Biochem.*, **49**, (1980).

Manganese

[4] G. D. Lawrence and D. T. Sawyer, 'The Chemistry of Biological Manganese', *Coord. Chem. Rev.*, **27**, 173 (1978).

Calcium

[5] R. H. Kretsinger and D. J. Nelson, 'Calcium in Biological Systems', *Coord. Chem. Rev.,* **18,** 29 (1976).

Molybdenum

[6] R. A. D. Wentworth, 'Mechanisms for the Reactions of Molybdenum in Enzymes', *Coord. Chem. Rev.,* **18,** 1 (1976).

[7] M. Coughlan (ed.), *Molybdenum and Molybdenum Containing Enzymes,* Pergamon Press, Oxford, 1980.

[8] K. B. Swedo and J. H. Enemark, 'Some Aspects of the Bioinorganic Chemistry of Molybdenum', *J. Chem. Ed.,* **56,** 70 (1979).

Methylmercury

[9] D. L. Rabenstein, 'The Aqueous Solution Chemistry of Methylmercury and its Complexes', *Acc. Chem. Res.,* **11,** 100 (1978).

Inorganic Drugs

Gold

[10] D. H. Brown and W. E. Smith, 'The Chemistry of the Gold Drugs used in the Treatment of Rheumatoid Arthritis', *Chem. Soc. Rev.,* **9,** 217 (1980).

Lithium

[11] F. N. Johnson, *Lithium Research and Therapy,* Academic Press, London, 1975.

[12] F. N. Johnson and S. Johnson, *Lithium in Medical Practice,* M.T. P. Press, Lancaster, 1978.

[13] S. Gershon and B. Shopsin, *Lithium, its Role in Psychiatric Research and Treatment,* Plenum Press, New York, 1973.

Platinum

[14] M. J. Cleare, 'Transition Metal Complexes in Cancer Chemotherpay', *Coord. Chem. Rev.,* **12,** 349 (1974).

Rhodium

[15] R. G. Hughes *et al., Am. Assoc. Cancer Res.,* **13,** 120 (1972). Rhodium carboxylates.

[16] R. D. Gillard *et al., Nature,* **223,** 735 (1969). Antibacterial activity.

Cancer

[17] D. R. Williams, 'Metals, Ligands and Cancer', *Chem. Rev.,* **72,** 203 (1972).

General

[18] D. D. Perrin, 'Medicinal Chemistry', *Topics in Current Chemistry 64: Inorganic Biochemistry,* Chapter 3, Springer Verlag, Berlin, 1976.

Index